Bombay Lectures On
HIGHEST WEIGHT
REPRESENTATIONS OF
INFINITE DIMENSIONAL
LIE ALGEBRAS

T0210429

ADVANCED SERIES IN MATHEMATICAL PHYSICS

Also in the Series

Volume 1: Mathematical Aspects of String Theory
             edited by S-T Yau (*Harvard*)

Advanced Series in Mathematical Physics
Vol. 2

Bombay Lectures On

HIGHEST WEIGHT REPRESENTATIONS OF INFINITE DIMENSIONAL LIE ALGEBRAS

by

V G Kac

(Massachusetts Institute of Technology, Cambridge)

A K Raina

(Tata Institute of Fundamental Research, Bombay)

 World Scientific

Singapore • New Jersey • Hong Kong

Published by

World Scientific Publishing Co. Pte. Ltd.
P.O. Box 128, Farrer Road, Singapore 9128

U. S. A. office: World Scientific Publishing Co., Inc.
687 Hartwell Street, Teaneck NJ 07666, USA

ISBN 9971-50-395-6
9971-50-396-4 (pbk)

Printed in Singapore by General Printing Services Pte. Ltd.

PREFACE

This book is a write-up of a series of lectures given by the first author at the Tata Institute, Bombay, during December 1985 — January 1986.

The dominant theme of these lectures is the idea of a highest weight representation. This idea goes through four different incarnations.

The first is the canonical commutation relations of the infinite-dimensional Heisenberg algebra (= oscillator algebra). Although this example is extremely simple, it not only contains the germs of the main features of the theory, but also serves as a basis for most of the constructions of representations of infinite-dimensional Lie algebras.

The second is the highest weight representations of the Lie algebra $g\ell_\infty$ of infinite matrices, along with their applications to the theory of soliton equations, discovered by Sato and Date-Jimbo-Kashiwara-Miwa. Here the main point is the isomorphism between the vertex and the "Dirac sea" realizations of the fundamental representations of $g\ell_\infty$, a kind of a Bose-Fermi correspondence.

The third is the unitary highest weight representations of the affine Kac-Moody (= current) algebras. Since there is now a book devoted to the theory of Kac-Moody algebras, it was decided to devote to them a minimum attention. In the lectures affine algebras play a prominent role only in the Sugawara construction as the main tool in the study of the fourth incarnation of the main idea, the theory of highest weight representations of the Virasoro algebra.

The main results of the representation theory of the Virasoro algebra which are proved in these lectures are the Kac determinant formula and the unitarity of the "discrete series" representations of Belavin-Polyakov-Zamolodchikov and Friedan-Qiu-Shenker.

We hope that this elementary introduction to the subject, written by a mathematician and a physicist, will prove useful to both mathematicians and physicists. To mathematicians, since it illustrates, on important examples, the interaction of the key ideas of the representation theory of infinite-dimensional Lie algebras; and to physicists, since this theory is turning before our very eyes into an important component of such domains of theoretical physics as soliton theory, theory of two-dimensional statistical models, and string theory.

Throughout the book, the base field is the field of complex numbers $\mathbb{C}$, unless otherwise stated, $\mathbb{R}$ denotes the set of real numbers, $\mathbb{Z}$ the set of integers, and $\mathbb{Z}_+$ (resp. $\mathbb{N}$) the set of non-negative (resp. positive) integers.

The authors wish to thank the participants of the lectures, especially S. M. Roy and S. R. Wadia, for valuable suggestions and comments.

CONTENTS

Preface   v

Lecture 1   1

1.1. The Lie algebra d of complex vector fields on the circle   1
1.2. Representations $V_{\alpha, \beta}$ of d 4
1.3. Central extensions of d: the Virasoro algebra   7

Lecture 2   11

2.1. Definition of positive-energy representations of Vir 11
2.2. Oscillator algebra $\mathscr{A}$ 12
2.3. Oscillator representations of Vir 15

Lecture 3   19

3.1. Complete reducibility of the oscillator representations of Vir 19
3.2. Highest weight representations of Vir 21
3.3. Verma representations $M(c, h)$ and irreducible highest weight representations $V(c, h)$ of Vir 23
3.4. More (unitary) oscillator representations of Vir 26

Lecture 4   33

4.1. Lie algebras of infinite matrices   33
4.2. Infinite wedge space F and the Dirac positron theory   35
4.3. Representation of GL_∞ and $g\ell_\infty$ in F. Unitarity of highest weight representations of $g\ell_\infty$ 38
4.4. Representation of a_∞ in F 43
4.5. Representations of Vir in F 45

Lecture 5   49

5.1. Boson-fermion correspondence   49
5.2. Wedging and contracting operators   51
5.3. Vertex operators. The first part of the boson-fermion correspondence   53
5.4. Vertex representations of $g\ell_\infty$ and a_∞ 56

Lecture 6   59
6.1. Schur polynomials   59
6.2. The second part of the boson-fermion correspondence   61
6.3. An application: structure of the Virasoro representations for $c = 1$ 64

Lecture 7   69
7.1. Orbit of the vacuum vector under GL_∞ 69
7.2. Defining equations for Ω in $F^{(0)}$ 69
7.3. Differential equations for Ω in $\mathbb{C}[x_1, x_2, \ldots]$ 71
7.4. Hirota's bilinear equations   72
7.5. KP hierarchy   74
7.6. N-soliton solutions   77

Lecture 8   81
8.1. Degenerate representations and the determinant $\det_n(c, h)$ of the contravariant form   81
8.2. The determinant $\det_n(c, h)$ as a polynomial in h 83
8.3. The Kac determinant formula   85
8.4. Some consequences of the determinant formula for unitarity and degeneracy   88

Lecture 9   93
9.1. Representations of loop algebras in $\overline{a}_\infty$ 93
9.2. Representations of $\widehat{g\ell}'_n$ in $F^{(m)}$ 96
9.3. The invariant bilinear form on $\widehat{g\ell}_n$. The action of $\widetilde{GL}_n$ on $\widehat{g\ell}_n$ 97
9.4. Reduction from a_∞ to $\widehat{s\ell}_n$ and unitarity of highest weight representations of $\widehat{s\ell}_n$ 100

Lecture 10   105
10.1. Nonabelian generalization of Virasoro operators: the Sugawara construction   105
10.2. The Goddard-Kent-Olive construction   113

Lecture 11   117
11.1. $\widehat{s\ell}_2$ and its Weyl group   117
11.2. The Weyl-Kac character formula and Jacobi-Riemann theta functions   119

11.3.   A character identity   124

Lecture 12   129
12.1.   Preliminaries on $\widehat{s\ell}_2$ 129
12.2.   A tensor product decomposition of some representations
        of $\widehat{s\ell}_2$ 130
12.3.   Construction and unitarity of the discrete series
        representations of *Vir*   132
12.4.   Completion of the proof of the Kac determinant
        formula   137
12.5.   On non-unitarity in the region $0 \leqslant c < 1, h \geqslant 0$ 138

References   141

Bombay Lectures On

HIGHEST WEIGHT REPRESENTATIONS OF INFINITE DIMENSIONAL LIE ALGEBRAS

LECTURE 1

In this series of lectures we shall demonstrate some basic concepts and methods of the representation theory of infinite dimensional Lie algebras on four main examples:

1. The oscillator algebra
2. The Virasoro algebra
3. The Lie algebra $g\ell_\infty$
4. The affine Kac-Moody algebras.

In fact, the interplay between these examples is one of the key methods of the theory.

1.1. The Lie algebra d of complex vector fields on the circle.

We shall first consider the Virasoro algebra, which is playing an increasingly important role in theoretical physics. It is also a natural algebra to consider from a mathematical point of view as it is a central extension of the complexification of the Lie algebra $Vect\ S^1$ of (real) vector fields on the circle S^1. We shall start by finding the structure of $Vect\ S^1$ and later consider its central extensions.

Any element of $Vect\ S^1$ is of the form $f(\theta)\ d/d\theta$, where $f(\theta)$ is a smooth real-valued function on S^1, with θ a parameter on S^1 and $f(\theta + 2\pi) = f(\theta)$. The Lie bracket of vector fields is:

$$[f(\theta)\frac{d}{d\theta}, g(\theta)\frac{d}{d\theta}] = (fg' - f'g)(\theta)\frac{d}{d\theta},$$

where prime stands for the derivative. A basis (over $\mathbb{R}$) for $Vect\ S^1$ is provided by the vector fields

$$\frac{d}{d\theta} \quad , \quad \cos{(n\theta)}\frac{d}{d\theta} \quad , \quad \sin{(n\theta)}\frac{d}{d\theta} \quad (n = 1, 2, \ldots) \ .$$

To avoid convergence questions we consider this as a vector space basis, so that $f(\theta)$, $g(\theta)$ are arbitrary trigonometric polynomials, and take its linear span over $\mathbb{C}$ as this permits us to introduce $\exp{(in\theta)}$ instead of $\cos{(n\theta)}$ and $\sin{(n\theta)}$. We thus obtain a complex Lie algebra, denoted by d, with a basis

$$d_n = i \exp{(in\theta)}\frac{d}{d\theta} = -z^{n+1}\frac{d}{dz} \quad (n \in \mathbb{Z}) \tag{1.1}$$

where $z = \exp{(i\theta)}$. These elements satisfy the following commutation relations:

$$[d_m, d_n] = (m - n)d_{m+n} \quad (m, n \in \mathbb{Z}) \ . \tag{1.2}$$

The Lie algebra $Vect\ S^1$ can be considered as the Lie algebra of the group G of orientation preserving diffeomorphisms of S^1. If ζ_1, ζ_2 are two elements of G then their product is defined by composition:

$$(\zeta_1 \cdot \zeta_2)(z) = \zeta_1(\zeta_2(z))$$

for each $z = \exp{(i\theta)}$ on S^1. If $f(z)$ is an element of the vector space of smooth complex-valued functions on S_1, then $\gamma \in G$ acts on $f(z)$ by

$$\pi(\gamma)f(z) = f(\gamma^{-1}(z)) \ . \tag{1.3}$$

This clearly defines a representation of G. We take γ close to the identity (as physicists do):

$$\gamma(z) = z(1 + \epsilon(z)) = z + \sum_{n=-\infty}^{\infty} \epsilon_n z^{n+1} \tag{1.4a}$$

where we have made a Laurent (or Fourier) expansion of $\epsilon(z)$ and the ϵ_n are to be retained up to first order only. Then

$$\gamma^{-1}(z) = z - \sum_{n=-\infty}^{\infty} \epsilon_n z^{n+1} \tag{1.4b}$$

and

$$\pi(\gamma) f(z) = f(z - \sum_n \epsilon_n z^{n+1}) = (1 + \sum_n \epsilon_n d_n) f(z) \qquad (1.5)$$

where the d_n are defined by (1.1). This shows that the d_n form a (topological) basis of the complexification of the Lie algebra of G.

In the following we shall consider the complex Lie algebra d and view $d \cap$ *Vect* S^1 as the subalgebra (over $\mathbb{R}$) of real elements. One way to do this is to regard $d \cap$ *Vect* S^1 as the subalgebra of fixed points for the operation of complex conjugation under which d_n maps to $-d_{-n}$ and a scalar λ to its complex conjugate $\bar{\lambda}$. It is more convenient, however, to introduce a slightly different operation defined by:

$$\omega(d_n) = d_{-n} \quad , \qquad (1.6a)$$

$$\omega(\lambda x) = \bar{\lambda}\omega(x) \quad , \qquad (1.6b)$$

so that

$$\omega([x , y]) = [\omega(y) , \omega(x)] \qquad (1.6c)$$

where $x, y \in d$, $\lambda \in \mathbb{C}$. Thus ω is an antilinear anti-involution having the algebraic properties of Hermitian conjugation. Now $d \cap$ *Vect* S^1 consists of elements of d fixed under $-\omega$.

The purely algebraic operation ω on d can become an adjoint operation with respect to a suitable scalar product if we have a representation of d in some vector space. Suppose that we have a unitary representation of the group G on a vector space V with a positive-definite Hermitian form $\langle \cdot \mid \cdot \rangle$. Identifying elements of G with corresponding operators, we have:

$$\langle g(u) \mid g(v) \rangle = \langle u \mid v \rangle \text{ for } g \in G , u, v \in V .$$

Going over to the Lie algebra, this means that

$$\langle x(u) \mid v \rangle = - \langle u \mid x(v) \rangle \text{ for } x \in Vect \ S^1 ,$$

and for any $x \in d$:

$$\langle x(u) \mid v \rangle = \langle u \mid \omega(x)(v) \rangle . \qquad (1.7)$$

This motivates the following definitions:

Definition 1.1. Let g be a Lie algebra and let ω be an antilinear anti-involution on g, i.e. an $\mathbb{R}$-linear involution satisfying (1.6b and c). Let V be a representation space of g and $\langle \cdot \mid \cdot \rangle$ an Hermitian form on V. We say that $\langle \cdot \mid \cdot \rangle$ is *contravariant* if (1.7) holds for all $x \in g$, and $u, v \in V$. When $\langle \cdot \mid \cdot \rangle$ is non-degenerate, this means that

$$x^{\dagger} = \omega(x) \text{ for all } x \in g . \qquad (1.7')$$

Here and further $x^{\dagger}$ stands for the Hermitian adjoint of the operator x. We further say that this representation is *unitary* if in addition

$$\langle v \mid v \rangle > 0 \text{ for all } v \in V, v \neq 0 .$$

1.2. Representations $V_{\alpha, \beta}$ of d.

We shall find representations of d by considering a suitable vector space on which the group G acts and determining the action of G in this space for elements close to the identity. Using (1.5) we shall determine the action of d_n in this vector space.

Let $V_{\alpha, \beta}$ denote that space of 'densities' of the form $P(z) z^{\alpha} (dz)^{\beta}$, where α and β are complex numbers and $P(z)$ is an arbitrary polynomial in z and z^{-1}. A basis for $V_{\alpha, \beta}$ is given by the set of vectors

$$v_k = z^{k+\alpha} (dz)^{\beta} \quad (k \in \mathbb{Z}) . \qquad (1.8)$$

From (1.3),

$$\pi(\gamma) v_k = (\gamma^{-1}(z))^{k+\alpha} (d\gamma^{-1}(z))^{\beta}$$

and if γ is of the form (1.4a), we can use (1.4b) for $\gamma^{-1}(z)$. Thus

$$\pi(\gamma) v_k = (z - \sum_n \epsilon_n z^{n+1})^{k+\alpha} ((1 - \sum_n \epsilon_n (n+1) z^n) dz)^{\beta}$$

$$= (1 - (k + \alpha) \sum_n \epsilon_n \, z^n) \, (1 - \beta \sum_n \epsilon_n (n + 1) z^n) \, z^{k+\alpha} (dz)^\beta$$

$$= (1 - \sum_n \epsilon_n (k + \alpha + \beta n + \beta) z^n) \, z^{k+\alpha} (dz)^\beta \ .$$

Comparing with (1.5) we see that

$$d_n(v_k) = -(k + \alpha + \beta + \beta n) v_{n+k} \quad (n, k \in \mathbb{Z}) \ . \tag{1.9}$$

Formula (1.9) defines a two-parameter family of representations of the Lie algebra d. Note from (1.9) that d_0 is diagonal:

$$d_0(v_k) = -(k + \alpha + \beta) v_k \ . \tag{1.10}$$

The operator d_0 is called the *energy operator*.

The following well-known lemma is useful in the proof of irreducibility:

Lemma 1.1. Let V be a representation of a Lie algebra g which decomposes as a direct sum of eigenspaces of a finite dimensional commutative subalgebra h:

$$V = \bigoplus_{\lambda \in h^*} V_\lambda \tag{1.11a}$$

where $V_\lambda = \{ v \in V \mid h(v) = \lambda(h)v \text{ for all } h \in h \}$, and h^* is the dual vector space of h. Then any subrepresentation U of V respects this decomposition in the sense that

$$U = \bigoplus_\lambda (U \cap V_\lambda) \ . \tag{1.12a}$$

Applying this to our case with $g = d$ and $h = \mathbb{C} \, d_0$ we obtain the following corollary:

Corollary 1.1. Let V be a representation of d which decomposes as a direct sum of eigenspaces of d_0:

$$V = \bigoplus_k V_k \tag{1.11b}$$

Then any subrepresentation U of V respects this decomposition:

$$U = \bigoplus_k (U \cap V_k) \tag{1.12b}$$

Proof. We first prove Corollary 1.1. Any $v \in V$ can be written in the form $v = \sum_{j=1}^{m} w_j$, where $w_j \in V_{\lambda_j}$ according to (1.11b), and $d_0(w_j) = \lambda_j w_j$, where $\lambda_j \neq \lambda_k$ for $j \neq k$ $(j, k = 1, \ldots, m)$. Then if $v \in U$, we have the following set of equations:

$$v = w_1 + w_2 + \ldots + w_m$$

$$d_0(v) = \lambda_1 w_1 + \lambda_2 w_2 + \ldots + \lambda_m w_m$$

$$d_0^2(v) = \lambda_1^2 w_1 + \lambda_2^2 w_2 + \ldots + \lambda_m^2 w_m$$

$$\cdots\cdots\cdots\cdots\cdots\cdots\cdots\cdots\cdots\cdots\cdots\cdots$$

$$d_0^{m-1}(v) = \lambda_1^{m-1} w_1 + \lambda_2^{m-1} w_2 + \ldots + \lambda_m^{m-1} w_m .$$

Since U is a subrepresentation of d, it follows that $v, d_0(v), \ldots, d_0^{m-1}(v)$ must all lie in U. We then have a system of equations with a matrix which is invertible since its determinant is a Vandermonde and hence not zero. Thus each of the w_j $(j = 1, \ldots, m)$ can be expressed as a linear combination of vectors in U, proving (1.12b). The proof of Lemma 1.1 is identical to that of the corollary, since we can always find $h \in h$ such that $\lambda_j(h) \neq \lambda_k(h)$ for $j \neq k$ and so we merely have to replace d_0 by this h in the above proof. ∎

Remark 1.1. The representation $V_{\alpha, \beta}$ is isomorphic to the representation $V_{\alpha+m, \beta}$ for $m \in \mathbb{Z}$, by the shift by m of indices of the basis.

Proposition 1.1. The representation $V_{\alpha, \beta}$ of d is reducible if (i) $\alpha \in \mathbb{Z}$ and $\beta = 0$, or (ii) $\alpha \in \mathbb{Z}$ and $\beta = 1$; otherwise it is irreducible.

Proof. Let U be a non-zero subrepresentation of $V_{\alpha, \beta}$. Then by Corollary 1.1, U is also a direct sum of some of the 1-dimensional subspaces $\mathbb{C} v_j$. Let $v_k \in U$. If U is not the whole of $V_{\alpha, \beta}$, there is at least one vector in $V_{\alpha, \beta} \setminus U$. Let us first suppose that there are at least two distinct vectors v_m and $v_n (m \neq n)$ in $V_{\alpha, \beta} \setminus U$. By (1.9),

$$d_{m-k}(v_k) = -(k + \alpha + \beta + \beta(m - k))v_m , \tag{1.13}$$

$$d_{n-k}(v_k) = - (k + \alpha + \beta + \beta(n - k))v_n \ . \qquad (1.14)$$

Since U is a representation of d, the right-hand sides of (1.13) and (1.14) must lie in U, which is only possible if

$$k + \alpha + \beta + \beta(m - k) = 0 = k + \alpha + \beta + \beta(n - k) \ .$$

Since $m \neq n$ we have $\beta = 0$ and so $k + \alpha = 0$. This shows that if U has codimension at least 2, then $\beta = 0$ and $\alpha \in \mathbb{Z}$. We may assume, from Remark 1.1, that $\alpha = 0$ so that $U = \mathbb{C}v_0$. Thus $V_{0,0}' = V_{0,0}/\mathbb{C}v_0$ is an irreducible representation of d.

Let us now suppose that U has codimension 1 and $v_m \in V_{\alpha,\beta} \backslash U$. Hence U has at least two vectors $v_k, v_\varrho (k \neq \varrho \neq m)$. Thus $d_{m-k}v_k$ and $d_{m-\varrho}v_\varrho$ lie in U, which implies by (1.9) that

$$k + \alpha + \beta(m - k + 1) = 0 = \varrho + \alpha + \beta(m - \varrho + 1) \ .$$

We thus get $\beta = 1$ and from Remark 1.1 we may assume that $\alpha = 0$, so that $m = - 1$ and $U = \{$ linear span of $v_k | k \in \mathbb{Z}, \ k \neq - 1 \}$. Thus, $V_{0,1}' = U$ is an irreducible representation of d. ∎

We put $V_{\alpha,\beta}' = V_{\alpha,\beta}$ if $V_{\alpha,\beta}$ is irreducible; otherwise, $V_{\alpha,\beta}'$ is as in the proof of Proposition 1.1, so that all $V_{\alpha,\beta}'$ are irreducible.

Only some of the irreducible representations $V_{\alpha,\beta}'$ can have a Hermitian contravariant form:

Proposition 1.2. Representation $V_{\alpha,\beta}'$ has a non-degenerate Hermitian contravariant form if and only if $\beta + \bar\beta = 1$ and $\alpha + \beta = \bar\alpha + \bar\beta$. All these representations are unitary.

Proof is straightforward. We leave it to the reader. ∎

1.3. Central extensions of d: the Virasoro algebra.

We shall now study Lie algebra extensions $\hat{d}$ of d by a 1-dimensional center $\mathbb{C}c$. This means that

$$\hat{d} = d \oplus \mathbb{C}c$$

and the commutation relations (1.2) are replaced by commutation relations of the form

$$[d_m, d_n] = (m - n)d_{m+n} + a(m, n)c ,$$

$$\quad (1.15)$$

$$[d_m, c] = 0 ,$$

where $a(m, n) \in \mathbb{C}$ and $m, n \in \mathbb{Z}$.

The function $a(m, n)$ cannot be arbitrary because of the anticommutativity of the bracket and the Jacobi identity.

We observe from (1.15) that putting $d_0' = d_0, d_n' = d_n - [a(0, n)/n]c$ $(n \neq 0)$ we have $[d_0', d_n'] = -nd_n'$ $(n \in \mathbb{Z})$. This transformation is merely a change of basis in d and so we can drop the primes and say that

$$[d_0, d_n] = -nd_n \ (n \in \mathbb{Z}) . \quad (1.16)$$

From the Jacobi identity for d_0, d_m, d_n we get

$$[d_0, [d_m, d_n]] = -(m + n)[d_m, d_n] . \quad (1.17)$$

Substituting (1.15) in (1.17) and using (1.16) we get $(m + n)a(m, n)c = 0$. This shows that $a(m, n) = \delta_{m, -n} a(m)$, so that (1.15) becomes:

$$[d_m, d_n] = (m - n)d_{m+n} + \delta_{m, -n} a(m)c \quad (m, n \in \mathbb{Z}) \quad (1.18)$$

where $a(m) = -a(-m)$ by anticommutativity. We now work out the Jacobi identity for d_ϱ, d_m, d_n with $\ell + m + n = 0$ using (1.18) and the antisymmetry of $a(m)$. We get

$$(m - n)a(m + n) - (2n + m)a(m) + (n + 2m)a(n) = 0 . \quad (1.19)$$

Putting $n = 1$ in (1.19) we get

$$(m - 1)a(m + 1) = (m + 2)a(m) - (2m + 1)a(1) . \quad (1.20)$$

Since $a(-m) = -a(m)$ we have $a(0) = 0$ and we have to solve (1.20) for positive values of m only. Equation (1.20) is a linear recursion relation and its space of solutions is at most 2-dimensional, since the knowledge of $a(1)$ and $a(2)$ gives all $a(m)$. We observe that $a(m) = m$ and $a(m) = m^3$ are solutions. Hence $a(m) = \alpha m + \beta m^3$ is the general solution. If $\beta = 0$ then by defining $d_0' = d_0 + \frac{1}{2} \alpha c, d_i' = d_i \ (i \neq 0)$, the $d_i' \ (i \in \mathbb{Z})$ span an algebra without central charge, i.e. d, so that $\bar{d}$ is a direct sum of Lie algebras d and $\mathbb{C}c$. Hence, for a nontrivial central extension, $\beta \neq 0$ while α can be chosen arbitrarily; we conventionally choose $\alpha = -\beta$ so that $a(m) = \beta(m^3 - m)$. By rescaling c, we can choose a fixed value for β. Conventionally we put $\beta = 1/12$. We thus arrive at the *Virasoro algebra*, the Lie algebra *Vir* with a basis

$$\{ d_m , m \in \mathbb{Z} ; c \}$$

and the following commutation relations

$$[d_m , c] = 0 , \tag{1.21a}$$

$$[d_m , d_n] = (m - n)d_{m+n} + \delta_{m, -n} \frac{(m^3 - m)}{12} c . \tag{1.21b}$$

Thus, we have proved the following

Proposition 1.3. Every non-trivial central extension of the Lie algebra d by a 1-dimensional center is isomorphic to the Virasoro algebra *Vir*. ∎

We have obtained along with the proof:

Corollary 1.2. If $[d_m , d_n] = (m - n)d_{m+n} + \delta_{m, -n} a(m)c$ defines a Lie algebra, then $a(m) = \alpha m + \beta m^3$ for some $\alpha, \beta \in \mathbb{C}$. ∎

In these lectures we shall be mostly concerned with the unitary representations of *Vir*. Unitarity is defined through a Hermitian contravariant form (Definition 1.1) and contravariance is defined in terms of an antilinear, antiinvolution ω of *Vir* defined by (cf. (1.6)):

$$\omega(d_n) = d_{-n} \quad (n \in \mathbb{Z}) , \tag{1.22a}$$

$$\omega(c) = c . \tag{1.22b}$$

In particular, d_0 and c are self-adjoint elements of *Vir*.

LECTURE 2

2.1. Definition of positive-energy representations of *Vir*.

In the previous lecture we introduced the Virasoro algebra

$$Vir = \mathbb{C}c + \sum_{n \in \mathbb{Z}} \mathbb{C}d_n \ ,$$

$$[c \, , d_n] \ = \ 0 \ , \tag{2.1}$$

$$[d_m \, , d_n] \ = \ (m - n)d_{m+n} \ + \ \delta_{m,-n} \, \frac{m^3 - m}{12} \, c \ ,$$

as a central extension of d, the Lie algebra of complex polynomial vector fields on S^1. The so-called "two cocycle" $a(m, n) = \delta_{m,-n} \, (m^3 - m)/12$ was probably first discovered by Gelfand and Fuchs [1968] who computed the entire cohomology ring of d. They showed that the algebra $H^*(d)$ is the tensor product of the algebra of polynomials generated by a single generator of degree 2 and an exterior algebra defined by a single generator of degree 3, and hence

$$H^j(d) \ = \ \mathbb{C} \ \text{ for } j \ = \ 0, 2, 3, \ldots , H^1(d) \ = \ 0 \ .$$

What we proved in the last lecture is actually that $H^2(d) = \mathbb{C}$. (The equality $H^1(d) = 0$ simply means that $[d, d] = d$.)

In the previous lecture we also constructed some representations of *Vir* with $c = 0$. As we can see from (1.10), in these representations the spectrum of d_0 is unbounded both from above and from below. We shall continue to use the terminology "energy operator" for d_0 as an element of *Vir*. This leads us naturally to define a 'positive-energy representation' of *Vir*:

Definition 2.1. A representation of *Vir* is called a *positive-energy* representation if d_0 is diagonal and all its eigenvalues are nonnegative.

We shall see that *Vir* has nontrivial positive-energy unitary representations only if $c \neq 0$ and this is one of the reasons why *Vir* is so much more interesting an algebra than d. We shall learn in Lecture 4 how to find positive-energy representations of *Vir* using Dirac's hole theory. Now, however, we consider the Virasoro [1970] construction of some unitary positive-energy representations of *Vir* in terms of bosonic oscillators.

2.2. Oscillator algebra $\mathscr{A}$.

Let $\mathscr{A}$ be the *oscillator* (Heisenberg) algebra, the complex Lie algebra with a basis $\{ a_n , n \in \mathbb{Z} ; \hbar \}$, and the commutation relations

$$
\begin{aligned}
[\hbar , a_n] &= 0 \, (n \in \mathbb{Z}) \, , \\
[a_m , a_n] &= m\delta_{m,-n} \, \hbar \, (m, n \in \mathbb{Z}) \, .
\end{aligned} \tag{2.2}
$$

We note that $[a_0 , a_n] = 0 \, (n \in \mathbb{Z})$ so that a_0 is a central element (zero mode).

Introduce the Fock space $B = \mathbb{C}[x_1, x_2, \ldots]$; this is the space of polynomials in infinitely many variables $x_1, x_2, \ldots$.

Given $\mu, \hbar \in \mathbb{R}$, define the following representation of $\mathscr{A}$ on B ($n \in \mathbb{N}$):

$$
a_n = \epsilon_n \, \partial/\partial x_n \, , \tag{2.3a}
$$

$$
a_{-n} = \hbar\epsilon_n^{-1} n \, x_n \, , \tag{2.3b}
$$

$$
a_0 = \mu I \, , \tag{2.3c}
$$

$$
\hbar = \hbar I \, . \tag{2.3d}
$$

It is clear that these operators satisfy (2.2). The ϵ_n are arbitrary real scale factors which will be useful later on. As such they are inessential, but this is not the case for the parameter μ.

Remark 2.1. In (2.3b, d) and further on, by abuse of notation we use the same symbol to denote an element of $\mathscr{A}$ on the left and its eigenvalue on the right.

Lemma 2.1. If $\hbar \neq 0$, then the representation (2.3) of $\mathscr{A}$ is irreducible.

Proof. Any polynomial in B can be reduced to a multiple of 1 by successive application of the a_n with $n > 0$ (the annihilation operators). Then successive application of the a_{-n} with $n > 0$ (creation operators) can give any other polynomial in B provided that $\hbar \neq 0$. ∎

The constant polynomial $v = 1$, which is called the *vacuum vector* of B, has the properties

$$a_n(v) = 0 \qquad \text{for } n > 0 , \tag{2.4a}$$

$$a_0(v) = \mu v , \tag{2.4b}$$

$$\hbar(v) = \hbar v . \tag{2.4c}$$

Proposition 2.1. Let V be a representation of $\mathscr{A}$ which admits a nonzero vector v satisfying (2.4) with $\hbar \neq 0$. Then monomials of the form $a_{-1}^{k_1} \ldots a_{-n}^{k_n}(v)$ $(k_i \in \mathbb{Z}_+)$ are linearly independent. If these monomials span V, then V is equivalent to the representation of $\mathscr{A}$ on B given by (2.3). In particular, this is the case if V is irreducible.

Proof. We have a mapping ϕ from B to V defined by $\phi(P(\ldots, x_n, \ldots)) = P(\ldots, (\epsilon_n/\hbar n)a_{-n}, \ldots)v$. It is clear that if P is an element of B, then $a_n(\phi(P)) = \phi(a_n(P))$, i.e. ϕ is an intertwining operator. Since B is irreducible, $\ker \phi = 0$ and so ϕ is an isomorphism if ϕ is onto. ∎

We define an antilinear anti-involution ω on $\mathscr{A}$ by

$$\omega(a_n) = a_{-n} \quad , \quad \omega(\hbar) = \hbar .$$

(Physicists use the notation $a_n^\dagger = a_{-n}$ instead).

Proposition 2.2. Let V be as in Proposition 2.1. Then V carries a unique Hermitian form $\langle \cdot \mid \cdot \rangle$ which is contravariant with respect to ω, and such that $\langle v \mid v \rangle = 1$ for the vacuum vector v. The distinct monomials $a_{-1}^{k_1} \ldots a_{-n}^{k_n}(v)$ $(k_i \in \mathbb{Z}_+)$ form an orthogonal basis with respect to $\langle \cdot \mid \cdot \rangle$. These monomials have norms given by

$$\langle a_{-1}^{k_1} \ldots a_{-n}^{k_n} v \mid a_{-1}^{k_1} \ldots a_{-n}^{k_n} v \rangle = \prod_{j=1}^{n} k_j! (\hbar_j)^{k_j} . \qquad (2.5)$$

Proof. If $\langle \cdot \mid \cdot \rangle$ is a contravariant Hermitian form, then both the orthogonality and (2.5) are proved by induction on $k_1 + \ldots + k_n$, proving uniqueness. One checks directly that the Hermitian form, for which monomials are orthogonal and have norms given by (2.5), is contravariant, proving existence. (A more conceptual proof of existence is given below.)   ∎

Corollary 2.1. The contravariant Hermitian form on V such that $\langle v \mid v \rangle = 1$ is positive-definite if and only if $\hbar > 0$.   ∎

Definition 2.2. Let P be an arbitrary polynomial in B. The *vacuum expectation value* of P, denoted by $\langle P \rangle$, is defined as the constant term of P.

One clearly has the following useful formula:

$$\langle \omega(P) \rangle = \overline{\langle P \rangle} . \qquad (2.6a)$$

We can now define for $P, Q \in B$:

$$\langle P \mid Q \rangle = \langle \omega(P) Q \rangle . \qquad (2.6b)$$

One checks using (2.6a) that this is a Hermitian form; it is obviously contravariant and $\langle 1 \mid 1 \rangle = 1$. Hence, by Proposition 2.2, formulas (2.5) and (2.6b) are equivalent.

Definition 2.3. Define the degree of the monomial $x_1^{j_1} \ldots x_k^{j_k}$ as $j_1 + 2j_2 + \ldots + kj_k$. Let B_j be the subspace of B spanned by monomials of degree j. B_j is clearly finite dimensional and $\dim B_j = p(j)$. Here and further $p(j)$ denotes the number of partitions of $j \in \mathbb{Z}_+$ into a sum of positive integers with $p(0) = 1$. We have

$$B = \bigoplus_{j \geqslant 0} B_j , \qquad (2.7)$$

the *principal gradation* of B. We define the q-*dimension* of B to be

$$\dim_q B = \sum_{j \geqslant 0} (\dim B_j) q^j .$$

Then we clearly have:

$$\dim_q B = \sum_{j \geq 0} p(j) q^j = 1/\varphi(q) \qquad (2.8a)$$

where

$$\varphi(q) = \prod_{j \in \mathbb{N}} (1 - q^j) . \qquad (2.8b)$$

Remark 2.2. The monomial $x_1^{j_1} \ldots x_k^{j_k}$ represents j_1 oscillators in state $1, j_2$ in state 2, etc. and hence the degree is essentially the energy of the state.

Putting $q = \exp(-\beta)$ we see that 'q-dimension' is related to the partition function of statistical mechanics.

2.3. Oscillator representations of *Vir*.

Our aim in introducing the oscillator algebra $\mathscr{A}$ and its Fock representation is to introduce the Virasoro operators L_k. These are defined in the Fock representation B with $\hbar = 1$ and $a_0 = \mu$ by:

$$L_k = \frac{1}{2} \sum_{j \in \mathbb{Z}} : a_{-j} \, a_{j+k} : \qquad (k \in \mathbb{Z}) , \qquad (2.9)$$

where the colons indicate 'normal ordering', defined by

$$: a_i a_j : \ = \ \begin{cases} a_i a_j & \text{if } i \leq j \\ \\ a_j a_i & \text{if } i > j . \end{cases} \qquad (2.10)$$

As a result of normal ordering, when the operator L_k is applied to any vector of B, only a finite number of terms in the sum contribute. Hence L_k makes sense in B. The fundamental property of the L_k is that they provide a representation of the Virasoro algebra *Vir* in B with central charge $c = 1$:

Proposition 2.3. The L_k satisfy the commutation relations

$$[L_m , L_n] = (m - n) L_{m+n} + \delta_{m, -n} \frac{(m^3 - m)}{12} . \qquad (2.11)$$

Thus the map $d_k \to L_k$ is a representation of the Virasoro algebra in B for $c = 1$. Moreover this representation is unitary.

Remark 2.3.   For general $\hbar > 0$, the operators $L_k / \hbar$ satisfy (2.11) as well.

Proof of Proposition 2.3.   From $\omega(a_n) = a_{-n}$ and the definition (2.9) it is easy to verify that $\omega(L_k) = L_{-k}$. Hence the Hermitian contravariant form defined in (2.6) is contravariant for *Vir*. Since the representation of $\mathscr{A}$ in B for $\hbar > 0$ is unitary, the same holds for the representation of *Vir* with $c = 1$. We therefore only have to verify (2.11).

Lemma 2.2. $[a_k, L_n] = k a_{k+n} \ (k, n \in \mathbb{Z}) \ .$ (2.12)

The following 'cutoff' procedure replaces calculations with infinite sums by calculations with finite sums.

Define the function ψ on $\mathbb{R}$ by:

$$\psi(x) = 1 \ \text{if} \ |x| \leqslant 1 \ ; \ \ \psi(x) = 0 \ \text{if} \ |x| > 1 \ . \qquad (2.13a)$$

Put

$$L_n(\epsilon) = \frac{1}{2} \sum_{j \in \mathbb{Z}} \ : a_{-j} \, a_{j+n} : \psi(\epsilon j) \ . \qquad (2.13b)$$

Note that $L_n(\epsilon)$ contains only a finite number of terms if $\epsilon \neq 0$ and that $L_n(\epsilon) \to L_n$ as $\epsilon \to 0$. More precisely, the latter statement means that, given $v \in B$, $L_n(\epsilon)(v) = L_n(v)$ for ϵ sufficiently small.)

Proof of Lemma 2.2. $L_n(\epsilon)$ as defined in (2.13b) differs from the same expression without normal ordering by a finite sum of scalars. This drops out of the commutator $[a_k, L_n(\epsilon)]$ and so

$$[a_k, L_n(\epsilon)] \ = \frac{1}{2} \sum_j [a_k, a_{-j} \, a_{j+n}] \, \psi(\epsilon j)$$

$$= \frac{1}{2} \sum_j [a_k, a_{-j}] a_{j+n} \, \psi(\epsilon j) + \frac{1}{2} \sum_j a_{-j} [a_k, a_{j+n}] \, \psi(\epsilon j)$$

$$= \frac{1}{2} k a_{k+n} \, \psi(\epsilon k) + \frac{1}{2} k a_{k+n} \, \psi(\epsilon(k + n))$$

using (2.2). The $\epsilon \to 0$ limit gives (2.12). ∎

End of proof of Proposition 2.3.

$$[L_m(\epsilon), \, L_n] = \frac{1}{2} \sum_j \, [a_{-j} \, a_{j+m} \, , \, L_n] \, \psi(\epsilon j)$$

$$= \frac{1}{2} \sum_j \, (-j) \, a_{n-j} \, a_{j+m} \, \psi(\epsilon j)$$

$$+ \frac{1}{2} \sum_j \, (j + m) \, a_{-j} \, a_{j+m+n} \, \psi(\epsilon j) \, .$$

We split the first sum into terms satisfying $j \geqslant (n - m)/2$ which are in normal order and reverse the order of terms for which $j < (n - m)/2$ using the commutation relations. In the same way we split the second sum into terms satisfying $j \geqslant -(m + n)/2$ and $j < -(m + n)/2$. Then,

$$[L_m(\epsilon), \, L_n] = \frac{1}{2} \sum_j \, (-j) : a_{n-j} \, a_{j+m} : \psi(\epsilon j)$$

$$+ \frac{1}{2} \sum_j \, (j + m) : a_{-j} \, a_{j+m+n} : \psi(\epsilon j)$$

$$- \frac{1}{2} \delta_{m, -n} \sum_{j=-1}^{-m} \, j(m + j) \, \psi(\epsilon j) \, .$$

Making the transformation $j \to j + n$ in the first sum and taking the limit $\epsilon \to 0$, we get (2.11) since $-\frac{1}{2} \sum_{j=-1}^{-m} \, j(m + j) = (m^3 - m)/12$. ∎

LECTURE 3

3.1 Complete reducibility of the oscillator representations of *Vir.*

In the previous lecture we found unitary representations of the Virasoro algebra
(2.1) with central charge $c = 1$ in terms of the operators L_k:

$$d_k \to L_k \ (k \in \mathbb{Z}), \quad c \to 1 \ . \tag{3.1}$$

The L_k defined by (2.9) can be written without the normal ordering notation as

$$L_k = \frac{\epsilon}{2} a_{k/2}^2 + \sum_{j > -k/2} a_{-j} \, a_{j+k} \ , \tag{3.2}$$

where $\epsilon = 0$ if k is odd, $= 1$ if k is even.

The operators a_k form a unitary irreducible representation of the oscillator
algebra $\mathscr{A}$ in the vector space B defined by (2.3). From (3.2) the 'energy operators' L_0 is given by

$$L_0 = \mu^2 /2 + \sum_{j>0} a_{-j} \, a_j \ . \tag{3.3}$$

From (2.7) we know that B can be written as a direct sum of finite dimensional
subspaces of elements of degree j. It is easily seen that each such subspace is an
eigenspace of L_0 with eigenvalue $\mu^2 /2 + j$, where $j \in \mathbb{Z}_+$. Thus

$$B = \bigoplus_{j \in \mathbb{Z}_+} B_j \ , \tag{3.4}$$

where B_j is the $(1/2\mu^2 + j)$-eigenspace of L_0 as well as the subspace of elements
of degree j.

The decomposition (3.4) of the representation space as a direct sum of finite-dimensional subspaces is extremely useful as it enables us to use algebraic methods in an infinite dimensional problem, just as in the previous lectures in the cases of d and $\mathscr{A}$ which have similar decompositions.

The decomposition (3.4) can now be used to discuss whether this representation of *Vir* in B is irreducible. For Lemma 1.1 applies to the present situation and we can conclude that any subrepresentation U of B will have the decomposition:

$$U = \bigoplus_{j \in \mathbb{Z}_+} (U \cap B_j) \ . \tag{3.5}$$

We recall from Proposition 2.3 that the representation of *Vir* in B is also unitary and $\omega(L_0) = L_0$. It follows that the eigenspaces of L_0 which appear in (3.4) are mutually orthogonal with respect to the Hermitian contravariant form $\langle \cdot \mid \cdot \rangle$ on B. Given a subrepresentation U of B with the decomposition (3.5), then, denoting by U_j the finite-dimensional vector space $U \cap B_j$, we can define a subspace $U^\perp$ by

$$U^\perp = \bigoplus_{j \in \mathbb{Z}_+} U_j^\perp \ , \tag{3.6}$$

where $U_j^\perp$ is the finite-dimensional orthogonal complement of U_j in B_j. Clearly we have:

$$B = U \oplus U^\perp \ , \tag{3.7}$$

since

$$U^\perp = \{ v \in B \mid \ \langle U \mid v \rangle = 0 \} \ . \tag{3.8}$$

It is now clear that $U^\perp$ is also an invariant subspace for *Vir*, since $\langle U \mid U^\perp \rangle = 0$ and $L_j U \subset U$ implies $0 = \langle L_j U \mid U^\perp \rangle = \langle U \mid L_{-j} U^\perp \rangle$ and so $U^\perp$ is also invariant under the L_j $(j \in \mathbb{Z})$. We have proved:

Proposition 3.1. The unitary representation (3.1) of the Virasoro algebra with central charge $c = 1$ in the Fock space B is a direct sum of irreducible representations. ∎

3.2. Highest weight representations of *Vir*.

We can construct a subrepresentation B' of B as follows. From the properties of the vacuum vector 1 listed in (2.4) and the definition (3.2) of the operators L_k, we see that

$$L_k(1) = 0 \qquad (k > 0) , \tag{3.9}$$

$$L_0(1) = h \cdot 1 , \quad \text{where } h = \mu^2/2 . \tag{3.10}$$

Let B' be the linear span of vectors

$$L_{-i_k} \ldots L_{-i_2} L_{-i_1}(v) , \tag{3.11}$$

where we take arbitrary finite sequences $0 < i_1 \leqslant i_2 \leqslant \ldots \leqslant i_k$. It is easy to see from (3.9), (3.10) and (3.11) that B' is invariant under the L_j and so we get a subrepresentation of B, called the *highest component* of B. Note that vectors (3.11) are not linearly independent in general.

Proposition 3.2. The representation in B' is an irreducible representation of *Vir*.

Proof. If B' is not irreducible, then, as above, it can be written as the direct sum of two representations: $B' = U \oplus U^\perp$. Each summand has a decomposition of the form (3.5). The vacuum vector 1 spans the h-eigenspace of L_0 and hence belongs to exactly one summand, which implies that all vectors of (3.11) are in this summand. Hence the other summand is 0 and B' is irreducible. ■

Definition 3.1. A *highest weight representation* of *Vir* is a representation in a vector space V which admits a nonzero vector v such that for given complex numbers c, h :

$$c(v) = cv , \tag{3.12a}$$

$$d_0(v) = hv , \tag{3.12b}$$

and V is the linear span of vectors of the form

$$d_{-i_k} \ldots d_{-i_1}(v) \quad (0 < i_1 \leqslant \ldots \leqslant i_k) . \tag{3.12c}$$

The pair (c, h) is called the *highest weight*, and v is called the *highest weight vector*.

Note that (3.12b, c) imply (see below):

$$d_i(v) = 0 \quad \text{for} \quad i > 0 . \tag{3.12d}$$

Remark 3.1. B' is an example of a highest weight representation of *Vir* with highest weight vector 1 and highest weight $(1, \mu^2/2)$. Note that V' is spanned by elements of the form (3.12c) since $L_n(1) = 0$ for $n > 0$ and since V' is irreducible.

Remark 3.2. It follows from the proof of Proposition 3.2 that a unitary highest weight representation of *Vir* is irreducible.

Remark 3.3. While B' is certainly an irreducible representation of *Vir* for $c = 1$, the question remains as to whether or not $B' = B$. The answer depends on the value of μ. We shall see that $B' = B$ for generic values of μ, but this does not hold for special values. For example, if $\mu = 0$ then the vector $u = a_{-1}(1)$ satisfies (3.9) and (3.10) with $h = 1$. Hence we can use u to form a highest weight representation of *Vir* with $c = 1$ which is orthogonal to B'.

Let V be a highest weight representation of *Vir* with highest weight (c, h). We observe that all vectors of the form (3.12c) with a fixed value of $j = i_1 + i_2 + \ldots + i_k$ span the eigenspace V_{h+j} of d_0 with eigenvalue $h + j$, so that we have:

$$V = \bigoplus_{j \in \mathbb{Z}_+} V_{h+j} . \tag{3.13}$$

It is clear that $\dim V_{h+j} \leqslant p(j)$ and that equality holds if and only if all such vectors are linearly independent. Note also that (3.12d) follows from (3.13).

The formal power series

$$\text{ch } V = \text{tr}_V q^{d_0} \equiv \sum_{j \in \mathbb{Z}_+} (\dim V_{h+j}) q^{h+j}$$

is called the *character* of the representation V of *Vir*.

3.3. Verma representations $M(c, h)$ and irreducible highest weight representations $V(c, h)$ of *Vir*.

Definition 3.2. If all the vectors of the form (3.12c) in a highest weight representation of *Vir* are linearly independent, then this highest weight representation is called a *Verma representation*.

We shall denote the Verma representation by $M(c, h)$, where c, h are defined by (3.12a, b).

Remark 3.4. We can conclude from Remark 3.3 that B' is a Verma representation for generic values of h, but not, in particular, for $h = 0$.

Since the Verma representation $M(c, h)$ is the highest weight representation for given c, h in which all vectors of the form (3.12c) are linearly independent, it follows that there is a homomorphism from $M(c, h)$ to any other highest weight representation of *Vir* with the same values of c, h which maps highest weight vector to highest weight vector and commutes with the action of *Vir*. This implies that any highest weight representation is isomorphic to a quotient of the Verma representation. It follows, in particular, that (c, h) determines the Verma representation $M(c, h)$ uniquely.

The existence of a Verma representation can be established for any c, h by standard Lie algebra techniques. We start with the universal enveloping algebra U of *Vir*, form the left ideal $I(c, h)$ generated by the elements $\{ d_n \, (n > 0),$ $d_0 - h \cdot 1, \ c - c \cdot 1 \}$ where 1 is the identity element of U, and let $M(c, h) = U/I(c, h)$. The algebra *Vir* acts on $M(c, h)$ by left multiplication so that $M(c, h)$ is a representation of *Vir*. Consider the identity element 1 of U and let v be its image in $M(c, h)$. Then for $n > 0$, $d_n(v) = 0$ and $c(v) = cv$, $d_0(v) = hv$. Thus $M(c, h)$ is a highest weight representation with highest weight (c, h). The linear independence of vectors of the form (3.12c) is assured by the Poincaré-Birkhoff-Witt theorem. Hence $M(c, h)$ is a Verma representation.

Proposition 3.3. (a) The Verma representation $M(c, h)$ has the decomposition

$$M(c, h) = \bigoplus_{k \in \mathbb{Z}_+} M(c, h)_{h+k} \tag{3.14}$$

where $M(c, h)_{h+k}$ is the $(h + k)$-eigenspace of d_0 of dimension $p(k)$ spanned by vectors of the form

$$d_{-i_s} \cdots d_{-i_1}(v) \text{ with } 0 < i_1 \leqslant \ldots \leqslant i_s, \ i_1 + \ldots + i_s = k \ .$$

One has:

$$\text{ch}\, M(c, h) = q^h/\varphi(q) , \qquad\qquad (3.15)$$

where $\varphi(q)$ is defined by (2.8b).

(b) $M(c, h)$ is indecomposable, i.e. we cannot find nontrivial subrepresentations V, W such that

$$M(c, h) = V \oplus W . \qquad\qquad (3.16)$$

(c) $M(c, h)$ has a unique maximal proper subrepresentation $J(c, h)$, and

$$V(c, h) = M(c, h)/J(c, h)$$

is the unique irreducible highest weight representation with highest weight (c, h). We have:

$$\text{ch}\, V(c, h) \leqslant q^h/\varphi(q) . \qquad\qquad (3.17)$$

Proof. (a) follows from the previous discussion.

(b) Lemma 1.1 applies so that, if (3.16) were true, both V and W would be graded according to (3.14) and therefore the vacuum vector v would belong to V or W. If v belongs to a subrepresentation, then this subrepresentation must coincide with $M(c, h)$.

(c) By Lemma 1.1 all proper subrepresentations are graded according to (3.14) and so is their sum which is also a proper subrepresentation since it does not contain v. The maximal subrepresentation $J(c, h)$ is therefore the sum of all proper subrepresentations. The rest follows immediately.   ∎

Note that the irreducible highest weight representation $V(c, h)$ of *Vir* can be defined as the unique irreducible representation having a vector v such that

$$d_i(v) = 0 \text{ for } i > 0, \quad d_0(v) = hv, \quad c(v) = cv . \qquad (3.18)$$

Similarly, the Verma representation $M(c, h)$ can be defined as the unique representation of *Vir* having a vector v satisfying (3.18) and such that all monomials (3.12c) are linearly independent.

Remark 3.5. Note that the representations $V(c, h)$ with $h \geqslant 0$ are precisely all irreducible positive energy representations of *Vir*.

The antilinear anti-involution ω of *Vir* extends to an antilinear anti-involution of its universal enveloping algebra U. From Proposition 3.3(a) it makes sense to define for each $u \in M(c, h)$ its *expectation value* $\langle u \rangle$ as the coefficient of the highest weight vector v in the expansion of u with respect to (3.14)(since $M(c, h)_h = \mathbb{C}v$). Since any element of U is a linear combination of elements of the form

$$R = (d_{-j_1} \cdots d_{-j_s})(d_0^k c^r)(d_{i_1} \cdots d_{i_t}) \ ,$$

where $s, t \geqslant 0$, $j_1 \geqslant \ldots \geqslant j_s > 0$, $i_1 \geqslant \ldots \geqslant i_t > 0$, and since

$$\omega(R) = (d_{-i_t} \cdots d_{-i_1})(d_0^k c^r)(d_{j_s} \cdots d_{j_1}) \ ,$$

we deduce the following important property of the expectation value:

$$\langle \omega(R)(v) \rangle = \langle \overline{R(v)} \rangle \text{ provided that } c, h \in \mathbb{R} \ . \tag{3.19}$$

Proposition 3.4. (a) Provided that c and h are real, $M(c, h)$ carries a unique contravariant Hermitian form $\langle \cdot | \cdot \rangle$ such that $\langle v | v \rangle = 1$, where v is the highest weight vector.
(b) The eigenspaces of d_0 are pairwise orthogonal.
(c) Ker $\langle \cdot | \cdot \rangle = J(c, h)$. Hence $V(c, h)$ carries a unique contravariant Hermitian form such that $\langle v | v \rangle = 1$, and this form is non-degenerate.

Proof. (a), (b) We define $\langle \cdot | \cdot \rangle$ on monomials $P(v) = d_{-i_m} \cdots d_{-i_1}(v)$ and $Q(v) = d_{-j_n} \cdots d_{-j_1}(v)$ by:

$$\langle P(v) | Q(v) \rangle = \langle \omega(P)Q(v) \rangle = \langle d_{i_1} \cdots d_{i_m} \ d_{-j_n} \cdots d_{-j_1}(v) \rangle \ . \tag{3.20}$$

This is a Hermitian form due to (3.19). This Hermitian form is obviously contravariant. Clearly (3.20) vanishes if $i_1 + \ldots + i_m \neq j_1 + \ldots + j_n$.
(c) Ker $\langle \cdot | \cdot \rangle \equiv \{u \in M(c, h) \ | \langle u | w \rangle = 0 \text{ for all } w \in M(c, h)\}$ clearly is a proper subrepresentation of $M(c, h)$ since the highest weight vector v is not contained in it. Moreover any proper subrepresentation V of $M(c, h)$ lies in Ker $\langle \cdot | \cdot \rangle$. This is clear from (3.20), since if $P(v) \in V$ and $Q(v) \in M(c, h)$,

then $\omega(Q)P(v) \in V$, and if $\langle \omega(Q)P(v) \rangle \neq 0$ we get, using Lemma 1.1, that $v \in V$ and hence $V = M(c, h)$. ∎

Proposition 3.5. There exists at most one unitary highest weight representation on *Vir* for a given highest weight (c, h), viz. $V(c, h)$. A necessary condition for the unitarity of $V(c, h)$ is that $c \geqslant 0, h \geqslant 0$.

Proof. The first statement follows immediately from Remark 3.2 and Proposition 3.4. A necessary condition for unitarity is that

$$c_n = \langle d_{-n}v \,|\, d_{-n}v \rangle \geqslant 0 \quad \text{for each} \ \ n \geqslant 0. \tag{3.20}$$

But contravariance and the commutation rules show that

$$c_n = 2nh + c(n^3 - n)/12 \ . \tag{3.21}$$

Putting $n = 1$, we get $c_1 = 2h$ so that we must have $h \geqslant 0$. Moreover, (3.21) shows that c_n is dominated by cn^3 for large n, so that $c \geqslant 0$ is also necessary. ∎

Remark 3.6. The Hilbert completion of every unitary representation $V(c, h)$ can be integrated to a projective representation of Diff S^1 (see Goodman-Wallach [1985]).

3.4. More (unitary) oscillator representations of *Vir*.

We have constructed a unitary representations of *Vir* for $c = 1, h \geqslant 0$ using the oscillator representation. Now if g is a Lie algebra with representations in V_1 and V_2, then it has a representation in $V_1 \otimes V_2$ defined by

$$a(v_1 \otimes v_2) = (av_1) \otimes v_2 + v_1 \otimes (av_2) \tag{3.22}$$

for every a in g. Moreover, if V_1 and V_2 are unitary, we can define a Hermitian form on $V_1 \otimes V_2$ by

$$\langle v_1 \otimes v_2 \,|\, w_1 \otimes w_2 \rangle = \langle v_1 \,|\, w_1 \rangle \ \langle v_2 \,|\, w_2 \rangle \tag{3.23}$$

and it is easy to verify that this is a positive-definite Hermitian contravariant form. Hence the tensor product of two unitary representations of g is unitary.

Taking tensor products of the oscillator representation with itself we can construct oscillator representations of *Vir* for any positive integral value of c and any $h \geqslant 0$. Taking the highest component, we have a unitary highest weight representation of *Vir* for any $c = 1, 2, \ldots$ and $h \geqslant 0$.

The following modification of the Virasoro construction was found by Fairlie (see Chodos-Thorn [1974]). Defining for arbitrary real λ, μ

$$\tilde{L}_0 = (\mu^2 + \lambda^2)/2 + \sum_{j > 0} a_{-j} a_j \tag{3.24a}$$

and, for $k \neq 0$

$$\tilde{L}_k = \frac{1}{2} \sum_{j \in \mathbb{Z}} a_{-j} a_{j+k} + i\lambda k a_k , \tag{3.24b}$$

we can easily verify, using (2.11) and (2.12), that

$$[\tilde{L}_m , \tilde{L}_n] = (m - n)\tilde{L}_{m+n} + \delta_{m,-n} \frac{(m^3 - m)}{12} (1 + 12\lambda^2) . \tag{3.25}$$

(As before, we take $\hbar = 1$ and $a_0 = \mu$.) Clearly $\omega(\tilde{L}_k) = \tilde{L}_{-k}$ and the form $\langle \cdot \mid \cdot \rangle$ defined on the Fock space B is a Hermitian contravariant form for this representation of *Vir* as well. Taking the highest component, we therefore have a unitary highest weight representation of *Vir* for $c = 1 + 12\lambda^2, h = (\lambda^2 + \mu^2)/2$. In the $c > 0, h > 0$ quadrant we therefore have unitarity of $V(c, h)$ for points (c, h) lying in the region between the line $c = 1$ and $h = (c - 1)/24$. Taking into account the possibility of tensoring, the situation is as summarized in Figure 3.1. The entire region to the right of the line $c = 1$ is a region of unitarity, with the exception of the shaded triangles. We shall see later that $V(c, h)$ is unitary here as well, but we do not have a manifestly unitary oscillator construction. (This is a very interesting open problem.)

The following notion is important in the analysis of irreducibility.

Definition 3.3. A vector u of a representation of *Vir* in V is called *singular* if it is non-zero and

$$d_n(u) = 0 \quad \text{for} \quad n > 0 . \tag{3.26}$$

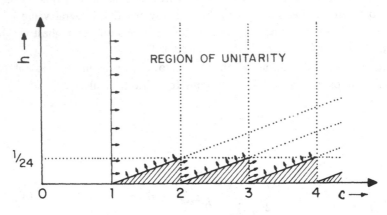

Figure 3.1

Of course a scalar multiple of the highest weight vector of a highest weight representation is singular. The following assertion is clear by Lemma 1.1 and Proposition 3.3.

Proposition 3.6.   A highest weight representation of *Vir* is irreducible if and only if it has no singular vectors other than scalar multiples of the highest weight vector.   ∎

We have not said anything about the region $0 < c < 1$, $h \geqslant 0$. We shall now show that there is an oscillator representation for the Virasoro algebra for $c = \frac{1}{2}$. The oscillators will, however, be fermionic oscillators ψ_n satisfying the anticommutation relations

$$[\psi_m , \psi_n]_+ \equiv \psi_m \psi_n + \psi_n \psi_m = \delta_{m,-n} \quad (m, n \in \delta + \mathbb{Z}) \;, \quad (3.27a).$$

where either $\delta = 0$ ('Ramond sector') or $\delta = \frac{1}{2}$ ('Neveu-Schwarz sector'). In the Ramond sector $(\delta = 0)$, $m = n = 0$ is allowed and we have

$$\psi_0^2 = \frac{1}{2} \;. \tag{3.27b}$$

The algebra (3.27) can be represented in the vector space

$$V_\delta = \Lambda[\xi_i \,|\, i \geqslant 0 \,, \, i \in \delta + \mathbb{Z}] \;, \tag{3.28}$$

where the symbol Λ on the right means the exterior algebra generated by the ξ_i. (Thus V_δ is the direct sum of $\mathbb{C}$ and antisymmetric tensors in the ξ_i of rank n for every $n \geqslant 1$. The antisymmetry is taken care of by the assumed relations

$$\xi_i \xi_j = -\xi_j \xi_i \qquad (3.29)$$

for all i, j.) The algebra (3.27) is represented in V_δ by the identifications

$$\left. \begin{array}{l} \psi_n \to \partial/\partial \xi_n \\[2ex] \psi_{-n} \to \xi_n \end{array} \right\} \qquad (n > 0) . \qquad (3.30)$$

For $n = 0$, which occurs in the Ramond sector only,

$$\psi_0 \to \frac{1}{\sqrt{2}} (\xi_0 + \partial/\partial \xi_0) . \qquad (3.31)$$

We shall identify the ψ_n with the corresponding operators in V_δ. The antilinear anti-involution ω, the vacuum expectation value $\langle \ \rangle$ and the Hermitian contravariant form $\langle \cdot \mid \cdot \rangle$ can be defined exactly as for the oscillator algebra. In fact, the monomials $\xi_{i_1} \ldots \xi_{i_s}$ ($i_1 < i_2 < \ldots < i_s$) form an orthonormal basis for $\langle \cdot \mid \cdot \rangle$, and therefore this representation is unitary.

Proposition 3.7. Let L_k ($k \in \mathbb{Z}$) be the operators in V_δ defined by

$$L_k = \delta_{k,0} \frac{(1 - 2\delta)}{16} + \frac{1}{2} \sum_j j : \psi_{-j} \psi_{j+k} : \qquad (3.32)$$

where j runs over $\delta + \mathbb{Z}$ and the normal ordering is defined by

$$\begin{array}{ll} : \psi_j \psi_k : \ = \ \psi_j \psi_k & \text{if } \ k \geqslant j \\[2ex] \qquad\quad = \ - \psi_k \psi_j & \text{if } \ k < j \ . \end{array} \qquad (3.33)$$

Then:

(i) $[\psi_m, L_k] = (m + k/2) \psi_{m+k} , \qquad (3.34)$

(ii) $[L_m , L_n] = (m - n)L_{m+n} + \delta_{m,-n} \dfrac{(m^3 - m)}{24}$.   (3.35)

Proof. A straighforward calculation using the same method as for the oscillator algebra in Lecture 2. ∎

From (3.32) we see that

$$L_0 = (1 - 2\delta)/16 + \sum_{j>0} j \, \psi_{-j}\psi_j \quad (j \in \delta + \mathbb{Z}) .$$ (3.36)

The representation of *Vir* in V_δ has central charge $c = \frac{1}{2}$. It is not irreducible: the subspaces V_δ^+ and V_δ^- of even and odd elements respectively are subrepresentations of V_δ. The elements of lowest energy in $V_{1/2}^+$ and $V_{1/2}^-$ are 1 and $\xi_{1/2}$ with energy 0 and $\frac{1}{2}$ respectively, and in V_0^+ and V_0^- are 1 and ξ_0, both of energy 1/16 (as is easily seen from (3.36)).

Furthermore, all four representations $V_\delta^\pm$ are irreducible. Indeed, by Proposition 3.6, if it were not the case, $V_\delta^\pm$ would contain a singular vector u of energy $h_0 + n_0$ where $h_0 = 0$, $\frac{1}{2}$ or 1/16 and $n_0 \geqslant 1$. The vector u would generate a unitary representation of *Vir* with highest weight $(\frac{1}{2}, h + n_0)$. But, as we shall see by the end of these lectures, the representations $V(\frac{1}{2}, h)$ are unitary for $h = 0$, 1/16 and $\frac{1}{2}$ only. Thus, we arrive at the following

Proposition 3.8. The representation of *Vir* given in Proposition 3.7 is irreducible in the even and odd subspaces $V_\delta^\pm$. In the Neveu-Schwarz sector we have a unitary highest weight representation of *Vir* with $c = \frac{1}{2}$, $h = 0$ in $V_{1/2}^+$ and with $c = \frac{1}{2}$, $h = \frac{1}{2}$ in $V_{1/2}^-$. In the Ramond sector we have a unitary highest weight representation of *Vir* with $c = \frac{1}{2}, h = 1/16$ in both V_0^+ and V_0^-. Finally we have:

$$\operatorname{ch} V\left(\frac{1}{2}, 0\right) \pm \operatorname{ch} V\left(\frac{1}{2}, \frac{1}{2}\right) = \prod_{n \in \mathbb{Z}_+} (1 \pm q^{n+1/2}) ,$$ (3.37)

$$\operatorname{ch} V\left(\frac{1}{2}, \frac{1}{16}\right) = q^{1/16} \prod_{n \in \mathbb{Z}_+} (1 + q^{n+1}) .$$ (3.38)

∎

So far we have constructed two types of irreducible representations of *Vir*: the representations $V'_{\alpha,\beta}$ and the highest weight representations $V(c, h)$. Of course, we also have the lowest weight representations $V^*(c, h)$, dual to $V(c, h)$.

Conjecture (Kac [1982]). An irreducible representation of *Vir* for which the energy operator is diagonalizable with finite dimensional eigenspaces is either $V'_{\alpha,\beta}$ or $V(c,h)$ or $V^*(c,h)$.

This conjecture has been checked by Kaplansky and Santharoubane [1985] in the case when all eigenspaces of d_0 have dimension $\leqslant 1$. Recently Chari and Pressley [1987] proved that the conjecture is true for unitary representations.

LECTURE 4

4.1. Lie algebras of infinite matrices.

In this lecture we shall study Lie algebras of infinite matrices and realize the algebras discussed earlier as its subalgebras. We shall then see how Dirac's positron theory can be given a representation-theoretic interpretation and used to obtain highest weight (= positive energy) representations of these Lie algebras.

Let

$$V = \bigoplus_{j \in \mathbb{Z}} \mathbb{C} v_j \tag{4.1}$$

be an infinite-dimensional complex vector space with a fixed basis $\{ v_j \mid j \in \mathbb{Z} \}$. We shall identify v_j with the column vector with 1 as the j-th energy and 0 elsewhere. Any vector in V has only a finite, but arbitrary, number of nonzero coordinates $(a_i)_{i \in \mathbb{Z}}$; this identifies V with $\mathbb{C}^\infty$, the space of such column vectors.

The Lie algebra $g\ell_\infty$ is defined by:

$$g\ell_\infty = \{ (a_{ij})_{i,j \in \mathbb{Z}} \mid \text{all but a finite number of the } a_{ij} \text{ are } 0 \} , \tag{4.2}$$

with the Lie bracket being the ordinary matrix commutator.

We denote by E_{ij} the matrix with 1 as the (i,j) entry and all other entries 0. The $E_{ij} (i,j \in \mathbb{Z})$ form a basis for $g\ell_\infty$. Clearly,

$$E_{ij} v_k = \delta_{jk} v_i \tag{4.3a}$$

and

$$E_{ij} E_{mn} = \delta_{jm} E_{in} . \tag{4.3b}$$

The commutation relations of $g\ell_\infty$ can be expressed as the commutation relations of the E_{ij}:

$$[E_{ij}, E_{mn}] = \delta_{jm} E_{in} - \delta_{ni} E_{mj} . \qquad (4.4)$$

The Lie algebra $g\ell_\infty$ can be viewed as the Lie algebra of the group GL_∞ defined as follows:

$$GL_\infty = \{ A = (a_{ij})_{i,j \in \mathbb{Z}} \mid A \text{ invertible and all but a} \qquad (4.5)$$

$$\text{finite number of } a_{ij} - \delta_{ij} \text{ are } 0 \} .$$

The group operation is matrix multiplication.

We define a bigger Lie algebra $\bar{a}_\infty$:

$$\bar{a}_\infty = \{ (a_{ij}) \mid i,j \in \mathbb{Z}, a_{ij} = 0 \text{ for } |i - j| \gg 0 \} . \qquad (4.6)$$

Matrices in $\bar{a}_\infty$ have a finite number of nonzero diagonals. It is easy to see that the product of two matrices in $\bar{a}_\infty$ is well defined, and is again in $\bar{a}_\infty$, so that $\bar{a}_\infty$ is a Lie algebra with the matrix commutator, containing $g\ell_\infty$ as a subalgebra.

We define the shift operators Λ_k by

$$\Lambda_k v_j = v_{j-k} . \qquad (4.7)$$

Clearly, by (4.3a),

$$\Lambda_k = \sum_{i \in \mathbb{Z}} E_{i, i+k} . \qquad (4.8)$$

Λ_k is the matrix with 1 at each entry on the k-th diagonal ($k = 0$ being the principal diagonal) and 0 elsewhere. The Λ_k form a commutative subalgebra of $\bar{a}_\infty$:

$$[\Lambda_j, \Lambda_k] = 0 \quad (j,k \in \mathbb{Z}) . \qquad (4.9)$$

In Lecture 1 we found representations of d in $V_{\alpha,\beta}$ given by equation (1.9). We shall change our notation by replacing v_k in (1.9) by v_{-k} so that (1.9) becomes:

$$d_n(v_k) = (k - \alpha - \beta(n + 1)) v_{k-n} , \qquad (4.10)$$

from which we deduce that

$$d_n = \sum_{k \in \mathbb{Z}} (k - \alpha - \beta(n + 1)) E_{k-n,k} \ . \tag{4.11}$$

Clearly the d_n are in $\overline{a}_\infty$ since d_n has nonzero entries only on the n-th diagonal. This gives an inclusion of d as a subalgebra of $\overline{a}_\infty$.

4.2. Infinite wedge space F and the Dirac positron theory.

In trying to construct a quantum theory of a single electron with positive energy, Dirac was led to construct a multiparticle theory of electrons and positrons. Let us recall the description given in his book 'The Theory of Quantum Mechanics' (Dirac [1958]):

"... the wave equation for the electron admits of twice as many solutions as it ought to, half of them referring to states with negative values for the kinetic energy ... we are led to infer that the negative-energy solutions ... refer to the motion of a new kind of particle having the mass of an electron and the opposite charge. Such particles have been observed experimentally and are called positrons. ... We assume that nearly all the negative-energy states are occupied, with one electron in each state in accordance with the exclusion principle of Pauli. An unoccupied negative-energy state will now appear as something with a positive energy, since to make it disappear, i.e. to fill it up, we should have to add to it an electron with negative energy. We assume that these unoccupied negative-energy states are the positrons.

These assumptions require there to be a distribution of electrons of infinite density everywhere in the world. A perfect vacuum is a region where all the states of positive energy are unoccupied and all those of negative energy are occupied ... the infinite distribution of negative-energy electrons does not contribute to the electric field. ... there will be a contribution -e for each occupied state of positive energy and a contribution +e for each unoccupied state of negative energy.

The exclusion principle will operate to prevent a positive-energy electron ordinarily from making transitions to states of negative energy. It will still be possible, however, for such an electron to drop into an unoccupied state of negative energy. In this case we should have an electron and positron disappearing simultaneously, their energy being emitted in the form of radiation. The converse process would consist in the creation of an electron and a positron from electromagnetic radiation."

An electron is described by its vector space of states, which we shall take to be the vector space V defined in (4.1). We shall call v_j the state of an electron of energy $j \in \mathbb{Z}$. The energy is thus not always positive. To fix this, Dirac theory requires us to consider an infinite number of such electrons satisfying the Pauli exclusion principle. We must therefore consider the *infinite wedge space* $\Lambda^\infty V$, where the symbol Λ stands for the exterior product, i.e., the antisymmetric tensor product. We can now define the vacuum state in accordance with Dirac theory as the state with positive energy states empty, but all negative energy states occupied. Denoting by $\wedge$ the exterior product of vectors, we have the perfect vacuum:

$$\psi_0 = v_0 \wedge v_{-1} \wedge v_{-2} \wedge \cdots . \qquad (4.12)$$

Note that we have included the zero-energy state with the negative energy states. All states are now produced by finite excitations of the perfect vacuum which produce simultaneously an electron with positive energy and a hole, i.e. an unoccupied state, in the negative-energy "sea". We define the state space

$$F^{(0)} = \Lambda^\infty_{(0)} V \qquad (4.13)$$

as the vector space with basis consisting of elements of the form

$$\psi = v_{i_0} \wedge v_{i_{-1}} \wedge v_{i_{-2}} \wedge \cdots , \qquad (4.14)$$

where

(i) $i_0 > i_{-1} > \ldots$ \hfill (4.15a)

(ii) $i_k = k$ for $k \ll 0$. \hfill (4.15b)

Conditions (4.15) ensure that ψ has an equal number of electrons and holes (positrons) so that ψ is a (charge conserving) excitation of the vacuum state ψ_0. We can compute the degree of the excitation of ψ when its labels satisfy (4.15a) by subtracting from each label of ψ the corresponding label of ψ_0. This leads us to define the *degree* of ψ (or the *energy*) as

$$\deg \psi = \sum_{s=0}^{\infty} (i_{-s} + s) . \qquad (4.16)$$

The degree of each ψ in $F^{(0)}$ is a finite non-negative integer because of (4.15b). Now let k be an arbitrary positive integer and let $\{k_0, k_1, \ldots, k_{n-1}\}$ be a partition of k in non-increasing order, i.e.

$$k = k_0 + k_1 + \ldots + k_{n-1} \qquad (4.17a)$$

where

$$k_0 \geqslant k_1 \geqslant \ldots \geqslant k_{n-1} . \qquad (4.17b)$$

Then this partition of k defines a unique ψ satisfying (4.15), *viz.*

$$\psi = v_{i_0} \wedge v_{j_{-1}} \wedge \cdots \wedge v_{j_{-n+1}} \wedge v_{-n} \wedge v_{-n-1} \wedge \cdots \qquad (4.18a)$$

where

$$j_{-i} = k_i - i \quad (i = 0, \ldots, n-1) . \qquad (4.18b)$$

This leads immediately to the following proposition:

Proposition 4.1. Let $F_k^{(0)}$ denote the linear span of all vectors of degree k. Then

(a) $\quad F^{(0)} = \underset{k \in \mathbb{Z}_+}{\oplus} F_k^{(0)} \; , \quad F_0^{(0)} = \mathbb{C}\psi_0 \qquad (4.19)$

(b) $\quad \dim F_k^{(0)} = p(k) \qquad (4.20)$

(c) $\quad \dim_q F^{(0)} \equiv \sum_k (\dim F_k^{(0)})q^k = 1/\varphi(q) \; , \qquad (4.21)$

where $\varphi(q)$ is defined by (2.8b).

Remark 4.1. An alternative way of computing deg ψ is by the Dirac recipe:

$$\deg \psi = \sum_s (i_s > 0 \text{ which occur}) - \sum_s (i_s \leqslant 0 \text{ which do not occur}) . \qquad (4.22)$$

The space $F^{(0)}$ is constructed starting from a particular reference vector, the perfect vacuum ψ_0. We can consider a larger vector space $F = \Lambda^\infty V$ with the basis consisting of all elements of the form (4.14) with labels satisfying (4.15a), with the sole restriction of Dirac theory that there should be only a finite number of unoccupied negative energy states (holes). Such elements will be called *semi-infinite monomials*. F can be decomposed as the vector sum of subspaces $F^{(m)}$:

$$F = \underset{m \in \mathbb{Z}}{\oplus} F^{(m)} \tag{4.23}$$

as follows. Each $F^{(m)}$ is based on a reference vector (vacuum)

$$\psi_m = v_m \wedge v_{m-1} \wedge v_{m-2} \wedge \cdots . \tag{4.24}$$

The space $F^{(m)}$ is the linear span of semi-infinite monomials of the form

$$\psi = v_{i_m} \wedge v_{i_{m-1}} \wedge \cdots , \tag{4.25}$$

where

(i) $i_m > i_{m-1} > \cdots$ $\tag{4.26a}$

(ii) $i_k = k + m$ for $k \ll 0$. $\tag{4.26b}$

We can define

$$\deg \psi = \sum_{s=0}^{\infty} (i_{m-s} + s - m) \tag{4.27}$$

$$= \sum_s (i_s > m \ \text{which occur}) - \sum_s (i_s \leqslant m \ \text{which do not occur}) . \tag{4.28}$$

Physicists call m the *charge number*.

Corollary 4.1. Each $F^{(m)}$ $(m \in \mathbb{Z})$ has a decomposition into subspaces $F_k^{(m)}$ of fixed degree $k \in \mathbb{Z}_+$ as in (4.19). The dimensions of these subspaces satisfy (4.20) and (4.21).

4.3. Representations of GL_∞ and $g\ell_\infty$ in F. Unitarity of highest weight representations of $g\ell_\infty$.

We can define representations R of GL_∞ and r of $g\ell_\infty$ in F by:

$$R(A)(v_{i_1} \wedge v_{i_2} \wedge \cdots) = Av_{i_1} \wedge Av_{i_2} \wedge \cdots , \tag{4.29}$$

$$r(a)(v_{i_1} \wedge v_{i_2} \wedge \ldots) = a v_{i_1} \wedge v_{i_2} \wedge \ldots$$

$$+ v_{i_1} \wedge a v_{i_2} \wedge \ldots + \ldots . \tag{4.30}$$

Equations (4.29) and (4.30) are related by:

$$\exp(r(a)) = R(\exp a), \quad a \in g\ell_\infty . \tag{4.31}$$

By (4.30) the basis elements of $g\ell_\infty$, *viz*. the E_{ij}, are represented by $r(E_{ij})$, where

$$r(E_{ij})v_{i_1} \wedge v_{i_2} \wedge \ldots = 0 \quad \text{if } j \notin \{ i_1, i_2, \ldots \} ,$$

$$\tag{4.32}$$

and $\quad = v_{i_1} \wedge \ldots \wedge v_{i_{k-1}} \wedge v_i \wedge v_{i_{k+1}} \wedge \ldots \quad \text{if } j = i_k .$

Of course the right-hand side of (4.32) vanishes if the label i is repeated. The $r(E_{ij})$ obey (4.4) and map each $F^{(m)}$ into itself so that the representation r of $g\ell_\infty$ in F is a direct sum of representations r_m in each $F^{(m)}$. Recall that $F^{(m)}$ is the linear span of semi-infinite monomials of the form

$$\psi = v_{i_m} \wedge \ldots \wedge v_{i_{m-k}} \wedge v_{m-k-1} \wedge \ldots , \tag{4.33a}$$

which, using (4.32), is

$$\psi = r(E_{i_m, m}) \ldots r(E_{i_{m-k}, m-k}) \psi_m . \tag{4.33b}$$

Define a positive definite Hermitian form $\langle \cdot \mid \cdot \rangle$ on F by declaring semi-infinite monomials to be an orthonormal basis. Let ω be the standard antilinear anti-involution of $g\ell_\infty$:

$$\omega(a) = a^\dagger , \tag{4.34}$$

where $a^\dagger$ denotes the transpose complex conjugate of the matrix a. One checks immediately (on the E_{ij}) that

$$\langle r(a)\psi \mid \psi' \rangle = \langle \psi \mid r(a^\dagger)\psi' \rangle , \tag{4.35}$$

so that the form $\langle \cdot \mid \cdot \rangle$ is contravariant and the representation r of $g\ell_\infty$ on F is unitary. The decomposition (4.23) is clearly orthogonal. Moreover, it follows now from (4.33b) (by the argument proving Proposition 3.2) that all representations r_m are irreducible. Thus, we have proved

Proposition 4.2.   The representation r of $g\ell_\infty$ in F is a direct sum of irreducible unitary representations r_m in $F^{(m)}$.   ■

The representation r was constructed by Kac and Peterson [1981] in a more general framework and was called the *infinite wedge representation*.

Each $F^{(m)}$ has a vector space decomposition into subspaces of fixed degree as noted in Corollary 4.1:

$$F^{(m)} = \bigoplus_{k \geqslant 0} F_k^{(m)} \ . \tag{4.36}$$

We can determine the action of $g\ell_\infty$ on this decomposition by examining the action of its basis element E_{ij} given in (4.32), where we see that E_{ij} either replaces v_j by v_i or gives zero. The replacement of v_j by v_i changes the degree of the vector by $i - j$. Thus

$$r(E_{ij}) F_k^{(m)} \subset F_{k+i-j}^{(m)} \ . \tag{4.37}$$

If we define

$$\deg E_{ij} = i - j \ ,$$

we can decompose $g\ell_\infty$ as the vector sum of homogeneous components g_j of degree j:

$$g\ell_\infty = \bigoplus_{j \in \mathbb{Z}} g_j \ . \tag{4.38}$$

A matrix in g_j has non-zero entries only on the $|j|$-th diagonal above $(j < 0)$ or below $(j > 0)$ the principal diagonal. By (4.37):

$$r(g_j) F_k^{(m)} \subset F_{k+j}^{(m)} \ . \tag{4.39}$$

Moreover,

$$r(g_j) \psi_m = 0 \text{ for } j < 0 \ , \tag{4.40}$$

since

$$r(E_{ij}) \psi_m = 0 \text{ for } i < j \ . \ \ .$$

Using (4.33b), we have:

$$F_k^{(m)} = \sum_{\substack{j_1 + \ldots + j_n = k \\ j_1, \ldots, j_n \in \mathbb{Z}_+}} r_m(g_{j_1}) \ldots r_m(g_{j_n}) \psi_m .$$ (4.41)

The vector space decomposition (4.38) is called the *principal gradation* of $g\ell_\infty$. The alternative definition (4.41) of the gradation of $F^{(m)}$ gives us a representation-theoretic interpretation of Dirac's definition of energy.

Let n_+ be the subalgebra of $g\ell_\infty$ consisting of strictly upper triangular matrices. Clearly,

$$n_+ = \bigoplus_{j<0} g_j .$$ (4.42)

Then from (4.40) and (4.42):

$$r_m(n_+) \psi_m = 0 ,$$ (4.43a)

$$r_m(E_{ii}) \psi_m = \lambda_i \psi_m ,$$ (4.43b)

where

$$\lambda_i = 1 \quad \text{if} \quad i \leqslant m$$
$$= 0 \quad \text{if} \quad i > m .$$ (4.44)

Definition 4.1. Given a collection of numbers $\lambda = \{ \lambda_i \mid i \in \mathbb{Z} \}$, called a *highest weight*, we define the *irreducible highest weight representation* π_λ of the Lie algebra $g\ell_\infty$ as an irreducible representation on a vector space $L(\lambda)$ which admits a non-zero vector v_λ, called a highest weight vector, such that

$$\pi_\lambda(n_+)v_\lambda = 0 ,$$ (4.45)

$$\pi_\lambda(E_{ii})v_\lambda = \lambda_i v_\lambda .$$ (4.46)

Note the analogy of this definition with the definition (3.18) of an irreducible highest weight representation of *Vir*. In particular, the same argument as in Lecture 3 shows that $L(\lambda)$ is determined by λ.

Thus for each $m \in \mathbb{Z}$ we have constructed an irreducible highest weight representation r_m of $g\ell_\infty$ with highest weight

$$\omega_m = \{ \lambda_i = 1 \text{ for } i \leqslant m , \lambda_i = 0 \text{ for } i > m \} . \qquad (4.47)$$

The r_m are called the *fundamental representations* of $g\ell_\infty$ and the ω_m the *fundamental weights*. Thus F is a direct sum of all fundamental representations of $g\ell_\infty$.

We showed in Lecture 3 (see equations (3.22), (3.23)) that the tensor product of two unitary representations V_1 and V_2 is also unitary. Furthermore, if V_1 and V_2 are irreducible unitary highest weight representations with highest weight vectors v_1 and v_2, then the vector $v_1 \otimes v_2$ is a highest weight vector of an irreducible subrepresentation of $V_1 \otimes V_2$, its *highest component*. It has highest weight equal to the sum of the two highest weights. Thus, we have proved the following proposition.

Proposition 4.3. The irreducible highest weight representations of $g\ell_\infty$ with highest weight of the form $\sum_i k_i \omega_i$, where the k_i are nonnegative integers, are unitary. ∎

It is easy to see that the unitarity of a highest weight representation of $g\ell_\infty$ with highest weight $\sum k_i \omega_i$ forces the k_i to be nonnegative integers (cf. Lecture 9).

Before concluding this subsection, let us note the following formula for the representation R_m of $A \in GL_\infty$ on $F^{(m)}$:

$$R_m(A) (v_{i_m} \wedge v_{i_{m-1}} \wedge \ldots) = \sum_{j_m > j_{m-1} > \ldots} \left(\det A_{j_m, j_{m-1}, \ldots}^{i_m, i_{m-1}, \ldots} \right)$$

$$\times v_{j_m} \wedge v_{j_{m-1}} \wedge v_{j_{m-2}} \wedge \ldots , \qquad (4.48)$$

where $A_{j_m, j_{m-1}, \ldots}^{i_m, i_{m-1}, \ldots}$ denotes the matrix located on the intersection of the rows $j_m, j_{m-1}, \ldots$ and columns $i_m, i_{m-1}, \ldots$ of the matrix A.

Proof. An immediate consequence of (4.29) and standard calculus of exterior algebra. ∎

4.4. Representation of a_∞ in F.

Matrices in $\bar{a}_\infty$ have a finite number of nonzero diagonals and so are finite linear combinations of matrices of the form

$$a_k = \sum_{i \in \mathbb{Z}} \lambda_i E_{i,\,i+k} \tag{4.49}$$

where the λ_i are arbitrary complex numbers. If we try to apply (4.30) to represent a_k in F, we find that for $k \neq 0, r(a_k)\, \psi_m$ is a finite linear combination of semi-infinite monomials in $F^{(m)}$, since the terms appearing in (4.49) vanish if $i + k > m$ or $i \leqslant m$. However, for $k = 0$ we get

$$r(a_0)\, \psi_m = (\lambda_m + \lambda_{m-1} + \dots)\, \psi_m \tag{4.50}$$

and the sum on the right-hand side can diverge. Since all vectors in $F^{(m)}$ are finite linear combinations of finite excitations of the vacuum vector ψ_m, we conclude that $r(a_k)$ can be defined by (4.30) for $k \neq 0$, but that this definition does not make sense for $k = 0$. We remove the "anomaly" of (4.50) by defining $\hat{r}_m$ by:

$$\hat{r}_m(E_{ij}) = r_m(E_{ij}) \quad \text{if} \quad i \neq j \quad \text{or} \quad i = j > 0\;, \tag{4.51a}$$

$$\hat{r}_m(E_{ii}) = r_m(E_{ii}) - I \quad \text{if} \quad i \leqslant 0\;. \tag{4.51b}$$

Replacing r_m by $\hat{r}_m$ in (4.50) we see that the right hand side is now a finite sum:

$$\hat{r}_m(a_0)\psi_m = \left(\sum_{i=1}^{m} \lambda_i \right) \psi_m \quad \text{if } m \geqslant 1,$$

$$= -\left(\sum_{i=0}^{m+1} \lambda_i \right) \psi_m \quad \text{if } m \leqslant -1 \text{ and } = 0 \text{ if } m = 0\;.$$

If $A \in \bar{a}_\infty$ then $\hat{r}_m(A)$ maps $F^{(m)}$ into itself. However, while the $r_m(E_{ij})$ obey the commutation rules (4.4), this is no longer the case for $\hat{r}_m(E_{ij})$. We can rewrite (4.4) as a set of four relations:

    (i) $[E_{ij}, E_{k\ell}] = 0$ for $j \neq k,\, \ell \neq i$

    (ii) $[E_{ij}, E_{j\ell}] = E_{i\ell}$ for $\ell \neq i$

    (iii) $[E_{ij}, E_{ki}] = -E_{kj}$ for $j \neq k$

    (iv) $[E_{ij}, E_{ji}] = E_{ii} - E_{jj}\;.$ (4.52)

These relations are satisfied by $r_m(E_{ij})$. Since the presence of I in the commutators on the left-hand side of (4.52) will leave the left-hand side unchanged, it follows from (4.51) that the $\hat{r}_m(E_{ij})$ will satisfy the first three equations of (4.52). In the last equation we get

$$[\,\hat{r}_m(E_{ij})\,,\hat{r}_m(E_{ji})\,] \;=\; \tilde{r}_m(E_{ii}) \,-\, \hat{r}_m(E_{jj}) \,+\, \alpha(E_{ij}\,,E_{ji})\,I$$

where

$$\alpha(E_{ij}\,,E_{ji}) \;\;=\; -\,\alpha(E_{ji}\,,E_{ij}) \;=\; 1 \;\;\text{if}\;\; i \leqslant 0\,,\, j \geqslant 1\,,$$

$$\alpha(E_{ij}\,,E_{mn}) \;=\; 0 \quad\text{in all other cases.}$$
(4.53)

Thus

$$\hat{r}_m([\,E_{ij}\,,E_{k\ell}\,]) \;=\; [\,\hat{r}_m(E_{ij})\,,\hat{r}_m(E_{k\ell})\,] \,-\, \alpha(E_{ij}\,,E_{k\ell})\;.\qquad(4.54)$$

Extending $\hat{r}_m$ to $\overline{a}_\infty$ by linearity, we get a projective representation of $\overline{a}_\infty$ due to the presence of the scalar summand in (4.54). This can be made into a linear representation of the central extension of $\overline{a}_\infty$, *viz.* the Lie algebra a_∞ defined by

$$a_\infty \;=\; \overline{a}_\infty \,\oplus\, \mathbb{C}c \qquad\qquad(4.55)$$

with $\mathbb{C}c$ in the center and bracket

$$[\,a,b\,] \;=\; ab \,-\, ba \,+\, \alpha(a,b)c\;,\qquad\qquad(4.56)$$

where the *two-cocycle* $\alpha(a,b)$ is linear in each variable and defined on the E_{ij} by (4.53). Extending $\hat{r}_m$ from $\overline{a}_\infty$ to a_∞ by $\hat{r}_m(c) = 1$, we obtain a linear representation $\hat{r}_m$ of the Lie algebra a_∞ in $F^{(m)}$. It is clear that by extending ω to a_∞ by using (4.34) on $\overline{a}_\infty$ and by defining $\omega(c) = c$, the representations $\hat{r}_m$ of a_∞ are unitary as well. Moreover, one can show in a similar fashion that every unitary irreducible highest weight representation of $g\ell_\infty$ extends to a unitary irreducible representation of a_∞.

The algebra a_∞ was introduced by Kac and Peterson [1981] and independently by Date, Jimbo, Kashiwara and Miwa [1981].

Let us consider first the shift operators Λ_k which (see (4.9)) form a commutative subalgebra of $\overline{a}_\infty$.

Under $\hat{r}_m$ this algebra will become

$$[\hat{r}_m(\Lambda_n), \hat{r}_m(\Lambda_k)] = \alpha(\Lambda_n, \Lambda_k) I .$$

It is a straightforward computation to show that

$$\alpha(\Lambda_n, \Lambda_k) = n \, \delta_{n,-k} , \qquad (4.57a)$$

so that

$$[\hat{r}_m(\Lambda_n), \hat{r}_m(\Lambda_k)] = n \, \delta_{n,-k} . \qquad (4.57b)$$

Note also that

$$\hat{r}_m(\Lambda_0) = mI . \qquad (4.58)$$

Comparing with (2.2) we see that (4.57b) is simply the commutation relations of the oscillator algebra $\mathscr{A}$. Note that the antilinear anti-involution ω of a_∞ is consistent with that defined on $\mathscr{A}$ (see Proposition 2.2). Thus we have constructed "fermionic" unitary representations $\hat{r}_m$ of the oscillator algebra.

4.5. Representations of *Vir* in *F*.

We have seen that the algebra d can be represented as a two-parameter family of subalgebras of $\bar{a}_\infty$ (recall (4.10), (4.11)). Consequently the projective representation of d in $F^{(m)}$ under $\hat{r}_m$ must be a linear representation of the 1-dimensional central extension of d, which we have already determined to be the Virasoro algebra:

$$[\hat{r}(d_i), \hat{r}(d_j)] = (i - j)\hat{r}(d_{i+j}) + \alpha(d_i, d_j) .$$

The computation of the 2-cocycle is straightforward:

$$\alpha(d_i, d_j) = \sum_{k,\ell} (k - \alpha - \beta(i + 1))(\ell - \alpha - \beta(j + 1)) \alpha(E_{k-i,k}, E_{\ell-j,\ell})$$

$$= \delta_{i,-j} \sum_{k=1}^{i} (k - \alpha - \beta(i + 1))(k - i - \alpha + \beta(i - 1))$$

$$= \delta_{i,-j} \left(\frac{(i^3 - i)}{12} c_\beta + 2ih_0 \right) , \qquad (4.59)$$

where

$$c_\beta = -12\beta^2 + 12\beta - 2 \,, \quad h_m = \frac{1}{2}(\alpha - m)(\alpha + 2\beta - 1 - m) \,.$$
$$\tag{4.60}$$

Defining L_i in $F^{(m)}$ by:

$$L_i = \hat{r}(d_i) \text{ if } i \neq 0 \,,$$

$$L_0 = \hat{r}(d_0) + h_0 \,.$$
$$\tag{4.61}$$

we see that

$$[L_i, L_j] = (i - j)L_{i+j} + \delta_{i,-j} \frac{(i^3 - i)}{12} c_\beta \,.$$
$$\tag{4.62}$$

From (4.61) it follows that

$$L_i \psi_m = 0 \text{ for } i > 0 \,; \quad L_0 \psi_m = h_m \psi_m \,.$$
$$\tag{4.63}$$

We have thus obtained (cf. Feigin and Fuchs [1982]), a representation of the Virasoro algebra on $F^{(m)}$ with central charge c_β and with minimal eigenvalue h_m of the energy operator given by (4.60). Note that these representations of *Vir* are in general non-unitary since the antilinear anti-involution of a_∞ is not consistent with that of *Vir*.

Remark 4.2.   The following four cases are of special interest:

   1) $\beta = \frac{1}{2}$,
   2) $\beta = 0$,
   3) $\beta = 1$,
   4) $\beta = -1$, $\alpha = 1$.

These are the wedge representation of *Vir* over the representation of d on half-densities, functions, differential forms and vector fields (= the adjoint representation) respectively. In the first case, $c = 1$ and the wedge representation is manifestly unitary (since the underlying representation is unitary); in the second and the third case, $c = -2$; in the fourth case, $c = -26$. The last case is intimately related to the fact that 26 is the critical dimension of the bosonic string theory (see e.g. Feigin [1984], Frenkel-Garland-Zuckerman [1986]).

Remark 4.3. Note that $c_\beta = (6\beta^2 - 6\beta + 1)c_1$. On the other hand, the Chern class of the determinant line bundle λ_β of the vector bundle on the moduli space of algebraic curves, whose fiber over a curve C is the space of differentials of degree β on C, is expressed by the same formula via the Chern class of the Hodge line bundle λ_1 (Mumford's theorem). This coincidence has been explained recently by Arbarello-De Concini-Kac-Procesi [1987] by establishing a canonical isomorphism between the second cohomology of the Lie algebra of differential operators of degree $\leqslant 1$ and the second singular cohomology of the moduli space of quadruples (C, p, v, L), where C is a smooth genus g Riemann surface, p a point on C, v a non-zero tangent vector to C at p and L a degree $g - 1$ line bundle on C.

LECTURE 5

5.1. Boson-fermion correspondence.

In Lecture 4 we realized the oscillator algebra $\mathscr{A}$ as a subalgebra of a_∞ by shift operators Λ_k:

$$[\Lambda_j , \Lambda_k] = j \, \delta_{j,-k} \, c \; . \tag{5.1}$$

For the representation $\hat{r}_m$ on $F^{(m)}$ we have:

$$\hat{r}_m(\Lambda_k) \, \psi_m = 0 \quad \text{for} \quad k > 0 \; . \tag{5.2}$$

Let us consider all elements of $F^{(m)}$ of the form:

$$\hat{r}_m(\Lambda_{-k_s}) \ldots \hat{r}_m(\Lambda_{-k_1}) \, \psi_m \quad (0 < k_1 \leqslant \ldots \leqslant k_s) \; . \tag{5.3}$$

Due to Proposition 2.1, these vectors are linearly independent. All of them with $\sum_i k_i = k$ lie in $F_k^{(m)}$ due to (4.37), and they form a basis of $F_k^{(m)}$ since the dimension of $F_k^{(m)}$ is exactly $p(k)$ (Corollary 4.1).

We have thus obtained an irreducible representation of the algebra of bosonic oscillators $\mathscr{A}$ in the fermionic space $F^{(m)}$ which is isomorphic to the representation of $\mathscr{A}$ in the space B of polynomials in infinitely many variables described in Lecture 2 (see Proposition 2.1). Let σ_m denote this isomorphism:

$$\sigma_m : F^{(m)} \overset{\sim}{\to} B^{(m)} = \mathbb{C}[x_1 , x_2 , \ldots] \; . \tag{5.4}$$

Note that $\sigma_m (\mathbb{C}\psi_m) = \mathbb{C}$ (as these are the elements killed by all Λ_k , $k > 0$). We normalize σ_m by the condition

$$\psi_m \to 1 \; . \tag{5.5a}$$

For the transported representation $\hat{r}_m^B = \sigma_m \, \hat{r}_m \sigma_m^{-1}$ of $\mathcal{A}$ on $B^{(m)}$ we have $(k > 0)$:

$$\hat{r}_m^B(\Lambda_k) = \frac{\partial}{\partial x_k} \, ,$$

(5.5b)

$$\hat{r}_m^B(\Lambda_{-k}) = k \, x_k \, , \quad \hat{r}_m^B(\Lambda_0) = m \, .$$

We have put the label 'm' on B in (5.4) to indicate that it is the copy of B corresponding to $F^{(m)}$ for $m \in \mathbb{Z}$. Note that (5.5a and b) completely determine σ_m.

To the principal gradation of $F^{(m)}$ by subspaces $F_k^{(m)}$ of energy k, there corresponds the principal gradation of

$$B^{(m)} = \bigoplus_{k \in \mathbb{Z}_+} B_k^{(m)}$$

defined by

$$\deg(x_j) = j \, .$$

(5.6)

This follows from (4.39).

We have also the contravariant Hermitian form on $B^{(m)}$ transported from $F^{(m)}$ via σ_m, which satisfies

$$\langle 1 \mid 1 \rangle = 1 \quad \text{and} \quad \hat{r}_m^B(\Lambda_k)^\dagger = \hat{r}_m^B(\Lambda_{-k}) \, .$$

(5.7)

Note that (2.6) can be rewritten as follows:

$$\langle P \mid Q \rangle = \bar{P}\left(\frac{\partial}{\partial x_1}, \frac{1}{2}\frac{\partial}{\partial x_2}, \frac{1}{3}\frac{\partial}{\partial x_3}, \dots \right) Q(x) \Big|_{x=0} \, .$$

(5.8)

Here $\bar{P}$ means taking complex conjugates of all coefficients of the polynomial P. Due to Proposition 2.2, this is the transported Hermitian form.

There are two natural questions concerning the isomorphism σ_m, which can be regarded as an algebraic version of the *boson-fermion correspondence*:

1. The semi-infinite monomials $v_{i_m} \wedge v_{i_{m-1}} \wedge \dots$ of $F^{(m)}$ are mapped by σ_m to some polynomials in $B^{(m)}$. What are these polynomials?
2. How can the representation of the subalgebra $\mathcal{A}$ of a_∞ in $B^{(m)}$ be extended to the whole Lie algebra a_∞?

We defer the answer to the first question to Lecture 6 and take up the second question.

It turns out simpler to deal with F itself rather than with each $F^{(m)}$. Hence we define the direct sum of maps

$$\sigma = \bigoplus_{m \in \mathbb{Z}} \sigma_m \ , \tag{5.9a}$$

so that

$$\sigma : F = \bigoplus_{m \in \mathbb{Z}} F^{(m)} \to B \equiv \bigoplus_{m \in \mathbb{Z}} B^{(m)} \ . \tag{5.9b}$$

To keep track of the index m on the right-hand side of (5.9b), we introduce a new variable z and put

$$B^{(m)} = z^m \, \mathbb{C} \, [\, x_1, x_2, \ldots \,] \ . \tag{5.10}$$

Thus

$$\sigma(\psi_m) = z^m \tag{5.11}$$

and we can view B as the polynomial algebra in $x_1, x_2, \ldots$ and z, z^{-1} :

$$B = \mathbb{C} \, [\, x_1, x_2, \ldots ; z, z^{-1} \,] \ . \tag{5.12}$$

We let $r^B = \sigma \, r \, \sigma^{-1}$ (resp. $\hat{r}^B = \sigma \, \hat{r} \, \sigma^{-1}$) be the transported representation of $\mathcal{A}_\infty$ from F to B.

5.2. Wedging and contracting operators.

We proceed to define wedging and contracting operators in F. Recall from (4.1) that

$$V = \bigoplus_{j \in \mathbb{Z}} \mathbb{C} \, v_j \ .$$

Defining the linear functional v_j^* on V by

$$v_j^*(v_i) = \delta_{ij} \quad (i, j \in \mathbb{Z}) \ , \tag{5.13}$$

we can define the *restricted dual* of V:

$$V^* = \bigoplus_{i \in \mathbb{Z}} \mathbb{C} \, v_i^* \, . \qquad (5.14)$$

Vectors in V and V^* define operators on F as follows. Each $v \in V$ defines a *wedging operator* $\hat{v}$ on F by:

$$\hat{v}(v_{i_1} \wedge v_{i_2} \wedge \ldots) = v \wedge v_{i_1} \wedge v_{i_2} \wedge \ldots \, . \qquad (5.15)$$

Each f in V^* defines a *contracting operator* $\check{f}$ on F by:

$$\check{f}(v_{i_1} \wedge v_{i_2} \wedge \ldots) = f(v_{i_1}) v_{i_2} \wedge v_{i_3} \wedge v_{i_4} \wedge \cdots$$

$$- f(v_{i_2}) v_{i_1} \wedge v_{i_3} \wedge v_{i_4} \wedge \cdots \qquad (5.16)$$

$$+ f(v_{i_3}) v_{i_1} \wedge v_{i_2} \wedge v_{i_4} \wedge \cdots - \cdots \, .$$

Note that the operators $\hat{v}_i$ and $\check{v}_i^*$ are adjoint with respect to the contravariant Hermitian form $\langle \, \cdot \mid \cdot \, \rangle$. The operator $\hat{v}$ maps $F^{(m)}$ into $F^{(m+1)}$, while $\check{f}$ maps $F^{(m)}$ into $F^{(m-1)}$. It is easy to see from (4.32) that

$$\hat{r}(E_{ij}) = \hat{v}_i \, \check{v}_j^* \, . \qquad (5.17)$$

Thus

$$\hat{r}(\Lambda_k) = \sum_{i \in \mathbb{Z}} \hat{v}_i \, \check{v}_{i+k}^* \quad \text{for} \quad k \neq 0 \, , \qquad (5.18a)$$

$$\hat{r}(\Lambda_0) = \sum_{i > 0} \hat{v}_i \, \check{v}_i^* - \sum_{i \leqslant 0} \check{v}_i^* \, \hat{v}_i \, . \qquad (5.18b)$$

The operators $\{ \hat{v}_i, \check{v}_j^* \mid i, j \in \mathbb{Z} \}$ generate a Clifford algebra:

$$[\hat{v}_i \, , \, \hat{v}_j]_+ = 0 \, , \quad [\check{v}_i^*, \, \check{v}_j^*]_+ = 0 \, , \quad [\hat{v}_i \, , \, \check{v}_j^*]_+ = \delta_{ij} \, . \qquad (5.19)$$

(Here $[\, , \,]_+$ stands for the anticommutator: $[a \, , b]_+ = ab + ba$.) From (5.18a) and (5.19) it is straightforward to verify the following commutation relations, which hold for $j \neq 0$:

$$[\hat{r}(\Lambda_j) \, , \, \hat{v}_k] = \hat{v}_{k-j} \qquad (5.20a)$$

$$[\hat{r}(\Lambda_j),\, \check{v}_k^*\,] \;=\; -\,\check{v}_{k+j}^* \;. \tag{5.20b}$$

Our aim is to determine $r^B(E_{ij})$. From (5.17) it is clear that this can be achieved by transforming $\hat{v}_i$ and $\check{v}_j^*$ by σ.

Remark 5.1. Putting $|0> = \psi_0$ and noting that $\hat{v}_j\,|0> = 0$ for $j \leqslant 0$ and $\check{v}_j^*\,|0> = 0$ for $j > 0$, we obtain an isomorphism between the wedge representation of the Clifford algebra and its spin representation, which is more familiar to physicists.

5.3. Vertex operators. The first part of the boson-fermion correspondence.

Let us introduce the *generating series*

$$X(u) \;=\; \sum_{j\in\mathbb{Z}} u^j\,\hat{v}_j, \quad X^*(u) \;=\; \sum_{j\in\mathbb{Z}} u^{-j}\,\check{v}_j^* \;, \tag{5.21}$$

where u is a nonzero complex number. As we shall see, the introduction of the generating series for $\hat{v}_i$, $\check{v}_j^*$ simplifies the determination of their transforms under σ. Since $X(u)$ is defined by an infinite series, it maps each $F^{(m)}$ into the *formal completion* $\hat{F}^{(m+1)}$ of $F^{(m+1)}$ in which infinite sums of semi-infinite monomials are permitted. Similarly, $X^*(u)$ maps $F^{(m)}$ into $\hat{F}^{(m-1)}$. We define

$$\hat{F} \;=\; \bigoplus_{m\in\mathbb{Z}} \hat{F}^{(m)} \;.$$

The transported operators $\sigma X(u)\sigma^{-1}$ and $\sigma X^*(u)\,\sigma^{-1}$ map B into $\hat{B}$, where $\hat{B}$ is the space of formal power series in $x_1, x_2, \ldots$ and z, z^{-1}, which are polynomial in z and z^{-1}.

From (5.20a, b) we find that for $j \neq 0$:

$$[\hat{r}(\Lambda_j),\, X(u)\,] \;\;=\; u^j\, X(u) \qquad , \tag{5.22a}$$

$$[\hat{r}(\Lambda_j),\, X^*(u)\,] \;=\; -\,u^j\, X^*(u) \;. \tag{5.22b}$$

These equations hold in $\hat{F}$; under the isomorphism $\sigma : \hat{F} \overset{\sim}{\to} \hat{B}$ they will hold in $\hat{B}$ as well. We already know the transform of Λ_j, namely, for $j > 0$ we have:

$$\hat{r}^B(\Lambda_j) \;\;=\; \sigma\,\hat{r}(\Lambda_j)\,\sigma^{-1} \;=\; \partial/\partial x_j$$

$$\hat{r}^B(\Lambda_{-j}) = \sigma \hat{r}(\Lambda_{-j}) \sigma^{-1} = j x_j .$$ (5.23)

Defining the *vertex operators* $\Gamma(u)$, $\Gamma^*(u)$ by

$$\Gamma(u) = \sigma X(u) \sigma^{-1}$$

$$\Gamma^*(u) = \sigma X^*(u) \sigma^{-1}$$

we see from (5.22a) that $\Gamma(u)$ satisfies the commutation relations:

$$[\partial/\partial x_j, \Gamma(u)] = u^j \Gamma(u)$$ (5.24a)

$$[x_j, \Gamma(u)] = \frac{u^{-j}}{j} \Gamma(u) ,$$ (5.24b)

with corresponding equations for $\Gamma^*(u)$ coming from (5.22b). The two commutation relations (5.24a, b) suffice to determine $\Gamma(u)$ (up to a constant); similarly, $\Gamma^*(u)$ is also determined by its corresponding commutation relations:

Proposition 5.1. $\Gamma(u)$ and $\Gamma^*(u)$ have the following form on $\hat{B}^{(m)}$:

$$\Gamma(u)\Big|_{\hat{B}(m)} = u^{m+1} z \exp\left(\sum_{j \geqslant 1} u^j x_j \right) \exp\left(-\sum_{j \geqslant 1} \frac{u^{-j}}{j} \frac{\partial}{\partial x_j} \right)$$ (5.25a)

$$\Gamma^*(u)\Big|_{\hat{B}(m)} = u^{-m} z^{-1} \exp\left(-\sum_{j \geqslant 1} u^j x_j \right) \exp\left(\sum_{j \geqslant 1} \frac{u^{-j}}{j} \frac{\partial}{\partial x_j} \right) .$$ (5.25b)

Proof. The factor z has to be present on the right-hand side of (5.25a) since $\Gamma(u)$ maps $\hat{B}^{(m)}$ into $\hat{B}^{(m+1)}$. Now let T_u be the operator on $\hat{B}$ defined by

$$(T_u f) (x_1, x_2, \ldots) = f(x_1 + u^{-1}, x_2 + \frac{u^{-2}}{2}, \ldots, x_j + \frac{u^{-j}}{j}, \ldots) .$$

By Taylor's formula

$$T_u = \exp\left(\sum_{j \geqslant 1} \frac{u^{-j}}{j} \frac{\partial}{\partial x_j} \right) .$$ (5.26)

It is now easy to verify that

$$[x_j, \Gamma(u) T_u] = 0,$$ (5.27)

by using (5.24b) and the simple relation

$$[x_j, T_u] = -\frac{u^{-j}}{j} T_u.$$

From (5.27) we conclude that $\Gamma(u) T_u$ contains no differential part, i.e. that

$$\Gamma(u) = z \; f(x_1, x_2, \ldots) \exp\left(\sum_{j \geq 1} \frac{u^{-j}}{j} \frac{\partial}{\partial x_j}\right),$$

where $f(x_1, x_2, \ldots)$ has to be determined. Using (5.24a) and the relation

$$\left[\frac{\partial}{\partial x_i}, \exp\left(-\sum_{j \geq 1} u^j x_j\right)\right] = -u^i \exp\left(-\sum_{j \geq 1} u^j x_j\right),$$

we find that

$$\left[\frac{\partial}{\partial x_i}, \exp\left(-\sum_{j \geq 1} u^j x_j\right) \Gamma(u)\right] = 0.$$ (5.28)

We conclude from (5.28) that

$$\Gamma(u) = c_m(u) z \; \exp\left(\sum_{j \geq 1} u^j x_j\right) \left(\exp - \sum_{j \geq 1} \frac{u^{-j}}{j} \frac{\partial}{\partial x_j}\right).$$

We can determine $c_m(u)$ by noting that the coefficient of the vacuum vector ψ_{m+1} of $\hat{F}^{(m+1)}$ in the expansion of $X(u)\psi_m$ is u^{m+1}. This completes the proof of (5.25a). By a similar argument we get (5.25b). ■

Define the operator $R(u): \hat{B} \to \hat{B}$ by

$$R(u)f(x, z) = uzf(x, uz).$$ (5.29)

Thus if $f(x, z) = z^m g(x_1, x_2, \dots)$ then

$$R(u) f(x, z) = u^{m+1} z^{m+1} g(x_1, x_2, \dots) \ .$$

We can now write down the general form of $\Gamma(u)$ and $\Gamma^*(u)$ (cf. Date-Jimbo-Kashiwara-Miwa [1983] and Kac-Peterson [1986]):

Theorem 5.1.

$$\Gamma(u) = R(u) \exp\left(\sum_{j \geqslant 1} u^j x_j \right) \exp\left(-\sum_{j \geqslant 1} \frac{u^{-j}}{j} \frac{\partial}{\partial x_j} \right) \qquad (5.30a)$$

$$\Gamma^*(u) = R(u)^{-1} \exp\left(-\sum_{j \geqslant 1} u^j x_j \right) \exp\left(\sum_{j \geqslant 1} \frac{u^{-j}}{j} \frac{\partial}{\partial x_j} \right) . \ (5.30b)$$

$$\blacksquare$$

Remark 5.2.   Theorem 5.1 is a discrete counterpart of the Skyrme model (see Skyrme [1971]). The idea of its proof is taken from Kac-Kazhdan-Lepowsky-Wilson [1981].

5.4. Vertex representations of $g\ell_\infty$ and a_∞.

We can now determine the representation of $g\ell_\infty$ and a_∞ in $B^{(m)}$ via the isomorphism σ_m. The basic element is E_{ij} which is represented in $F^{(m)}$ by $\hat{v}_i \check{v}_j^*$. The preceding section has shown us that the transforms of $\hat{v}_i$ and $\check{v}_j^*$ are very complicated, but that it is easier to deal with their generating functions. We shall therefore consider the generating function

$$\sum_{i, j \in \mathbb{Z}} u^i v^{-j} E_{ij} \ . \qquad (5.31a)$$

The representation in $\hat{F}$ of this generating function under r is simply

$$X(u) \ X^*(v) \ . \qquad (5.31b)$$

It is a straightforward computation to show, using Theorem 5.1 and

$$(\exp \ a \ \partial/\partial x) \, (\exp \ bx) = (\exp \ ab) \, (\exp \ bx) \, (\exp \ a \ \partial/\partial x) \ ,$$

that we have

Proposition 5.2.

$$\sum_{i,j \in \mathbb{Z}} u^i v^{-j} \, r^B(E_{ij}) \equiv \sigma_m(X(u)X^*(v)) \, \sigma_m^{-1} = \frac{(u/v)^m}{1 - (v/u)} \, \Gamma(u, v) \qquad (5.32)$$

where $\Gamma(u, v)$ is the following *vertex operator*:

$$\Gamma(u, v) = \exp\left(\sum_{j \geqslant 1} (u^j - v^j) x_j \right) \exp\left(-\sum_{j \geqslant 1} \frac{u^{-j} - v^{-j}}{j} \frac{\partial}{\partial x_j} \right) \qquad (5.33)$$

and we have assumed that $|v/u| < 1$. ∎

In the case of $\hat{r}^B$ we observe from (4.51) that we must simply subtract

$$\sum_{i \leqslant 0} (u/v)^i = \left(1 - \frac{v}{u} \right)^{-1}$$

from the right-hand side of (5.32), where we have once more assumed that $|v/u| < 1$. Thus we arrive at the following:

Proposition 5.3.

$$\sum_{i,j} u^i v^{-j} \, \hat{r}_m^B(E_{ij}) = \frac{1}{1 - (v/u)} \left(\left(\frac{u}{v} \right)^m \Gamma(u, v) - 1 \right). \qquad (5.34)$$

∎

To calculate $r_m^B(E_{ij})$ or $\hat{r}_m^B(E_{ij})$ we have to determine the coefficient of $u^i v^{-j}$ on the right-hand sides of (5.32) and (5.34).

This *vertex representation* of a_∞ was discovered in the case $m = 0$ by Date-Jimbo-Kashiwara-Miwa [1981].

LECTURE 6

6.1. Schur polynomials.

In Lecture 5 we asked for the explicit form of the polynomials in the bosonic Fock space $B = \mathbb{C}[x_1, x_2, \ldots]$ which correspond under σ_m to the semi-infinite monomials of $F^{(m)}$ (recall that they form an orthonormal basis in $F^{(m)}$). To find their image in B we need to first introduce the Schur polynomials.

Definition 6.1. The elementary Schur polynomials $S_k(x)$ are polynomials belonging to $\mathbb{C}[x_1, x_2, \ldots]$ and are defined by the generating function

$$\sum_{k \in \mathbb{Z}} S_k(x) z^k = \exp \sum_{k=1}^{\infty} x_k z^k . \tag{6.1}$$

Thus

$$S_k(x) = 0 \text{ for } k < 0 , \quad S_0(x) = 1 , \tag{6.2a}$$

$$S_k(x) = \sum_{k_1 + 2k_2 + \ldots = k} \frac{x_1^{k_1}}{k_1!} \frac{x_2^{k_2}}{k_2!} \ldots \text{ for } k > 0 . \tag{6.2b}$$

In particular

$$S_1(x) = x_1 , \quad S_2(x) = x_1^2/2 + x_2 ,$$

$$S_3(x) = x_1^3/6 + x_1 x_2 + x_3 , \tag{6.3}$$

$$S_4(x) = x_1^4/24 + x_2^2/2 + x_1^2 x_2/2 + x_1 x_3 + x_4 .$$

The elementary Schur polynomials are related to the complete symmetric functions h_k, where h_k is the sum of all monomials of total degree k in the variables $\epsilon_1, \ldots, \epsilon_N$. The generating function for the h_k is

$$\sum_{k \geqslant 0} h_k \, z^k = \prod_{i=1}^{N} (1 - \epsilon_i z)^{-1} \, . \tag{6.4}$$

To see the connection with the elementary Schur polynomials, substitute

$$x_j = \frac{\epsilon_1^j + \ldots + \epsilon_N^j}{j} \tag{6.5}$$

in the right-hand side of (6.1). We find that this expression reduces to the right-hand side of (6.4), which means that

$$S_k(x) = h_k(\epsilon_1, \ldots, \epsilon_N) \, . \tag{6.6}$$

We shall denote the set of all partitions by *Par*. Thus $\lambda \in Par$ is a non-increasing finite sequence of positive integers $\{\lambda_1 \geqslant \lambda_2 \geqslant \ldots \geqslant \lambda_k > 0\}$.

Definition 6.2. To each $\lambda = \{\lambda_1 \geqslant \lambda_2 \geqslant \ldots \geqslant \lambda_k\} \in Par$ we associate the *Schur polynomial* $S_\lambda(x)$ defined by the $k \times k$ determinant

$$S_{\lambda_1, \lambda_2, \ldots}(x) = \begin{vmatrix} S_{\lambda_1} & S_{\lambda_1+1} & S_{\lambda_2+2} & \cdots \cdots \\ S_{\lambda_2-1} & S_{\lambda_2} & S_{\lambda_2+1} & \cdots \cdots \\ S_{\lambda_3-2} & S_{\lambda_3-1} & S_{\lambda_3} & \cdots \cdots \\ \cdots\cdots\cdots\cdots\cdots\cdots\cdots \\ \cdots\cdots\cdots\cdots\cdots\cdots\cdots \end{vmatrix} \tag{6.7}$$

$$= \det \left(S_{\lambda_i + j - i}(x) \right) \, .$$

Remark 6.1. It is well-known that

$$S_\lambda(x) = \text{tr}_{\pi_\lambda} \begin{bmatrix} \epsilon_1 & & & \\ & \ddots & & \\ & & \ddots & \\ & & & \epsilon_N \end{bmatrix}$$

where π_λ is the representation of GL_N corresponding to the partition λ and the ϵ_i are related to the x_j by (6.5) (see e.g. Macdonald [1979]). Formula (6.6) shows this in the case when π_λ is the k-th symmetric power of the natural representation of GL_N.

We find from (6.3) and (6.7):

$$S_{1,1} = x_1^2/2 - x_2 \ ,$$

$$S_{2,1} = x_1^3/3 - x_3 \ , \tag{6.8}$$

$$S_{2,2} = x_1^4/12 - x_1 x_3 + x_2^2 \ .$$

It is clear from (6.2) and (6.7) that, with respect to the principal gradation on B introduced in Lecture 2 (in which deg $x_j = j$), the Schur polynomial $S_{\lambda_1, \lambda_2, \ldots}(x)$ is a homogeneous polynomial of degree $|\lambda| = \lambda_1 + \lambda_2 + \ldots$.

6.2. The second part of the boson-fermion correspondence.

We can now state the second part of the boson-fermion correspondence. For simplicity we state the theorem for $m = 0$ (see Corollary 6.1 for the generalization to all m):

Theorem 6.1.

$$\sigma_0(v_{i_0} \wedge v_{i_{-1}} \wedge \cdots) = S_{i_0, i_{-1}+1, i_{-2}+2, \ldots}(x) \ , \tag{6.9}$$

where $i_0 > i_{-1} > \ldots$ and $i_{-k} = -k$ for k sufficiently large.

Proof. Our strategy will be to compute

$$\sigma_0 \{ R_0 \ (\exp(y_1 \Lambda_1 + y_2 \Lambda_2 + \ldots)) v_{i_0} \wedge v_{i_{-1}} \wedge \cdots \}$$
$$= R_0^B \ (\exp(y_1 \Lambda_1 + y_2 \Lambda_2 + \ldots)) P(x) \tag{6.10}$$

where

$$P(x) = \sigma_0(v_{i_0} \wedge v_{i_{-1}} \wedge \ldots) . \tag{6.11}$$

We shall obtain the result by comparing the coefficient of the vacuum on the two sides of (6.10), recalling from (5.5a) that $\sigma_0(\psi_0) = 1$.

Before we can proceed any further we must first settle a technical problem, since $\exp(y_1\Lambda_1 + y_2\Lambda_2 + \ldots)$ is clearly not in GL_∞ (the argument of the exponential is equally not in $g\ell_\infty$). We must consider instead the larger group $\overline{GL}_\infty$ defined by:

$$\overline{GL}_\infty = \{\, A = (a_{ij}) \mid i,j \in \mathbb{Z}, \quad A \text{ invertible and all but a finite}$$
$$\text{number of the } a_{ij} - \delta_{ij} \text{ with } i \geqslant j \text{ are } 0 \,\} . \tag{6.12}$$

Thus matrices in $\overline{GL}_\infty$ have only a finite number of nonzero elements below the principal diagonal and it is evident that matrix multiplication is well defined. The Lie algebra of $\overline{GL}_\infty$ is:

$$\overline{g\ell}_\infty = \{\, (a_{ij}) \mid i,j \in \mathbb{Z}, \quad \text{all but a finite number of the}$$
$$a_{ij} \text{ with } i \geqslant j \text{ are } 0 \,\} .$$

$\overline{GL}_\infty$ and $\overline{g\ell}_\infty$ act not on V, but on a completion $\overline{V}$ of V defined as

$$\overline{V} = \Big\{ \sum_j c_j v_j \mid c_j = 0 \quad \text{for } j \gg 0 \Big\} .$$

On the other hand, it is easy to see that the representations R and r extend to representations of $\overline{GL}_\infty$ and $\overline{g\ell}_\infty$ on the same space F constructed from V. In particular the formula (4.48) holds for $R_m(A)$ when $A \in \overline{GL}_\infty$. The exponential map is defined on the whole of $\overline{g\ell}_\infty$ and we have

$$\exp r(a) = R(\exp a) \quad \text{for } a \in \overline{g\ell}_\infty .$$

It is clear that if $a = y_1\Lambda_1 + y_2\Lambda_2 + \ldots$, then $a \in \overline{g\ell}_\infty$ and $\exp a \in \overline{GL}_\infty$. Hence from the above discussion it follows that (4.48) can be used for $R(\exp a)$. We can now proceed with the proof.

In the bosonic picture, $r_0(\Lambda_k)$ is represented by $\partial/\partial x_k$ for $k > 0$, so that

$$R_0^B(\exp(y_1\Lambda_1 + y_2\Lambda_2 + \ldots)) = \exp \sum_{j \geqslant 1} y_j \frac{\partial}{\partial x_j} .$$

Now let $F(y)$ denote the coefficient of 1 when this operator is applied to $P(x)$. Then

$$F(y) = \exp\left(\sum_{j \geqslant 1} y_j \frac{\partial}{\partial x_j}\right) P(x) \Bigg|_{x=0}$$

$$= P(x + y)\Bigg|_{x=0} = P(y) \, ,$$

i.e.

$$F(y) = P(y) \, . \tag{6.13}$$

Now, $\exp\left(\sum_{k \geqslant 1} \Lambda_k\, y_k\right) = \exp\left(\sum_{k \geqslant 1} \Lambda_1^k\, y_k\right) = \sum_{k \geqslant 0} \Lambda_k\, S_k(y)$, using (6.1). This latter expression can be regarded as a matrix A with matrix elements

$$A_{mn} = S_{n-m}(y) \quad (m, n \in \mathbb{Z}) \, . \tag{6.14a}$$

Recalling from (6.2) that $S_k(x) = 0$ for $k < 0$, we see that $A \in \overline{GL}_\infty$. Hence (6.10) reduces to:

$$\sigma_0(v_{i_0} \,\wedge\, v_{i_{-1}} \,\wedge\, \ldots) = \text{coefficient of } \psi_0 \text{ in the expansion of}$$

$$\sigma_0\{R(A)(v_{i_0} \,\wedge\, v_{i_{-1}} \,\wedge\, \ldots)\} \, .$$

We can read off the required coefficient from (4.48), which gives

$$\det\left(A^{i_0,\, i_{-1},\, i_{-2},\, \cdots}_{0,\, -1,\, -2,\, \ldots}\right) \, .$$

This expression is the determinant of the matrix of elements of A at the intersections of rows $0, -1, -2, \ldots$ and columns $i_0, i_{-1}, i_{-2}, \ldots$ of A. From (6.7) and (6.14a) this is easily seen to be $S_{i_0,\, i_{-1}+1,\, i_{-2}+2,\, \ldots}$ (y). Thus, we have

$$F(y) = S_{i_0,\, i_{-1}+1,\, \ldots}(y) \, . \tag{6.14b}$$

Comparing (6.13) and (6.14b) completes the proof. ∎

Corollary 6.1.

$$\sigma_m(v_{i_m} \wedge v_{i_{m-1}} \wedge \cdots) = S_{i_m - m, \, i_{m-1} - m+1, \ldots}(x) . \qquad (6.15)$$

Corollary 6.2. In the course of the proof we have determined the action $R_m^B(A)$ of $A \in \overline{GL}_\infty$ in $B^{(m)}$:

$$R_m^B(A) S_\lambda = \sum_{\mu \in Par} \det (A_{\mu_1 + m, \, \mu_2 + m - 1, \ldots}^{\lambda_1 + m, \, \lambda_2 + m - 1, \ldots}) S_\mu . \qquad (6.16)$$

Corollary 6.3. The Schur polynomials form an orthonormal basis in B with respect to the contravariant Hermitian form $\langle \cdot \mid \cdot \rangle$ (defined by (5.8)), i.e.,

$$\langle S_\lambda \mid S_\mu \rangle = \delta_{\lambda, \mu} . \qquad (6.17)$$

6.3. An application: structure of the Virasoro representations for $c = 1$.

In Lecture 2 we saw that the Virasoro operators

$$L_k = \frac{\epsilon}{2} a_{k/2}^2 + \sum_{j > -k/2} a_{-j} \, a_{j+k} , \qquad (6.18)$$

where $\epsilon = 0$ for k odd, $\epsilon = 1$ for k even, satisfy the Virasoro algebra for $c = 1$. The a_k have a representation in $\mathbb{C}[\, x_1, x_2, \ldots \,]$ given by (2.3) with $\hbar = 1$. By the isomorphism established between $\mathbb{C}[\, x_1, x_2, \ldots \,]$ and $F^{(0)}$ we can choose the following representation of the a_k for $k > 0, \mu \in \mathbb{R}$:

$$a_k = \sqrt{2} \, \hat{r}_0(\Lambda_k) , \quad a_{-k} = \frac{1}{\sqrt{2}} \hat{r}_0(\Lambda_{-k}) , \quad a_0 = \mu/\sqrt{2} . \qquad (6.19)$$

We have made use of the freedom in choosing the ϵ_n in (2.3).

From (6.18) we find that:

$$L_0^{(\mu)} = \mu^2/4 + \sum_{j \geqslant 1} \hat{r}_0(\Lambda_{-j}) \hat{r}_0(\Lambda_j) \qquad (6.20a)$$

$$L_1^{(\mu)} = \mu \hat{r}_0(\Lambda_1) + \sum_{j \geqslant 1} \hat{r}_0(\Lambda_{-j}) \hat{r}_0(\Lambda_{j+1}) \qquad (6.20b)$$

$$L_2^{(\mu)} = \mu \hat{r}_0(\Lambda_2) + \hat{r}_0(\Lambda_1)^2 + \sum_{j \geqslant 1} \hat{r}_0(\Lambda_{-j}) \hat{r}_0(\Lambda_{j+2}) . \qquad (6.20c)$$

This gives us a representation of the Virasoro algebra for $c = 1$ in $F^{(0)}$.

Recall the Definition 3.3 of a singular vector in a representation of *Vir*. Note that it is an immediate consequence of the Virasoro algebra that v is a singular vector if (3.26) holds for $j = 1$ and $j = 2$.

We saw in Lecture 3 that the oscillator representation of the Virasoro algebra for $c = 1$ is a direct sum of unitary, irreducible highest weight representations. Each such representation is generated from a singular vector, *viz.* a highest weight vector. As we shall see in a moment, for generic values of μ in (6.20) there is only one singular vector, *viz.* the vacuum vector ψ_0 of $F^{(0)}$, and this representation is a Verma representation. However, there are special values of μ for which this is not the case:

Lemma 6.1. Let $\mu = -m \in \mathbb{Z}$ and $k \in \mathbb{Z}_+$ be such that $k + m \geqslant 0$. Consider

$$f_{m,k} = v_{k+m} \wedge v_{k+m-1} \wedge \cdots \wedge v_{m+1} \wedge v_{-k} \wedge v_{-k-1} \wedge \cdots \in F^{(0)} . \tag{6.21}$$

Then

$$L_j^{(-m)} f_{m,k} = 0 \quad \text{for } j > 0 , \tag{6.22}$$

$$L_0^{(-m)} f_{m,k} = \frac{1}{4} (m + 2k)^2 f_{m,k} . \tag{6.23}$$

Proof. A direct and lengthy computation which we omit. ∎

In Lemma 6.1 we have identified for each $\mu = -m \in \mathbb{Z}$ an infinite sequence of singular vectors. There are two obvious questions. (i) Are there nontrivial singular vectors for other values of μ? (ii) Are there any singular vectors for $\mu = -m$ other than those listed in Lemma 6.1?

Proposition 6.1. (a) If $\mu \notin \mathbb{Z}$ then all singular vectors of *Vir* in $F^{(0)}$ are multiples of the vacuum vector ψ_0.
(b) If $\mu = -m \in \mathbb{Z}$, then any singular vector of *Vir* in $F^{(0)}$ is a linear combination of the $f_{m,k}$ with $k, k + m \in \mathbb{Z}_+$.

Proof. (a) This is an immediate consequence of the Kac determinant formula, which we shall discuss in Lecture 8. (Thus, $M(1, h)$ is irreducible for $h \neq m^2/4$ ($m \in \mathbb{Z}$), i.e. if $\mu \notin \mathbb{Z}$.)

(b) We shall not give the proof. A quick way to see the correctness of the result is as follows. Any singular vector is a linear combination of singular eigenvectors of $L_0^{(-m)}$. We see from (6.20a) that a singular eigenvector has a $L_0^{(-m)}$-eigenvalue of the form $h = m^2/4 + n$, where $n \in \mathbb{Z}_+$, and generates a subrepresentation $V(1, h)$. From the Kac determinant formula it follows that h is of the form $h = (m + j)^2/4, j \in \mathbb{Z}_+$. Comparing we see that j is even, i.e. $j = 2k$ and $n = k(k + m)$. Since $n, k \in \mathbb{Z}_+$, we have $k + m \in \mathbb{Z}_+$. These are the sole subrepresentations allowed and from Lemma 6.1 we see that we have a singular vector for each of them. Hence the space $F^{(0)}$ is a direct sum of irreducible representations $V(1, (m + 2k)^2/4)$ for $k, k + m \in \mathbb{Z}_+$ if each representation occurs only once. For the multiplicity question we may appeal to the results of Feigin and Fuchs [1983b] who have shown that the multiplicity is indeed 1. ∎

From Proposition 6.1(b) it follows that:

$$F^{(0)} = \underset{k \geqslant 0, \, k+m \geqslant 0}{\oplus} V\left(1, \frac{1}{4}(m + 2k)^2\right) . \qquad (6.24)$$

Recall that (see 4.21))

$$\dim_q F^{(0)} = \varphi(q)^{-1} .$$

The subspace $V(1, \tfrac{1}{4}(m + 2k)^2)$ is generated from $f_{m,k}$ which has degree $k^2 + mk$. Hence from (6.24) we obtain

$$\frac{1}{\varphi(q)} = \sum_{k \in \mathbb{Z}_+} \operatorname{ch} V\left(1, \frac{1}{4}(m + 2k)^2\right) q^{k^2 + mk} . \qquad (6.25)$$

Now $f_{m,0} = \psi_0$ so that $V(1, m^2/4)$ is generated from ψ_0. Its character will be lowered from the Verma representation value of $1/\varphi(q)$ by the presence of the subrepresentation $V(1, (m + 2)^2/4)$ generated by the singular vector $f_{m,1}$ of degree $m + 1$. Hence,

$$\operatorname{ch} V(1, m^2/4) \leqslant (1 - q^{m+1})/\varphi(q) . \qquad (6.26)$$

Comparing (6.25) and (6.26) we see that consistency requires that equality holds in (6.26).

Using the isomorphism between $F^{(0)}$ and $\mathbb{C}[x_1, x_2, \ldots]$ we can summarize the obtained results as follows:

Theorem 6.2. Consider the representation

$$d_k \to L_k^{(\mu)} \ , \ c \to 1$$

of the Virasoro algebra on the space $\mathbb{C}[x_1, x_2, \ldots]$. Then:

(a) If $\mu \notin \mathbb{Z}$, this representation is irreducible and hence is a Verma representation.

(b) Let $\mu = -m \in \mathbb{Z}$. Put

$$P_{m,k}(x) = \underbrace{S_{k+m, \ldots, k+m}(x)}_{k \text{ times}} \quad (k \geqslant 0, \ k+m \geqslant 0) \ .$$

Then the $P_{m,k}$ are singular vectors with eigenvalues $(m+2k)^2/4$ and all singular vectors are linear combinations of the $P_{m,k}$. Furthermore, we have:

$$\mathbb{C}[x_1, x_2, \ldots] = \underset{\substack{k \in \mathbb{Z}_+ \\ k \geqslant -m}}{\oplus} V(1, (m+2k)^2/4) \ .$$

(c) $\operatorname{ch} V(1, m^2/4) = (1 - q^{m+1})/\varphi(q)$. ∎

This theorem is due to several authors: Kac [1979], Segal [1981], Wakimoto-Yamada [1986]. The results of this subsection are not used in the sequel.

LECTURE 7

7.1. Orbit of the vacuum vector under GL_∞.

In Lecture 4 we constructed a representation of the group GL_∞ in $F^{(0)}$ and hence in $B = \bar{C}[x_1, x_2, \ldots]$ by the boson-fermion correspondence. We shall use this correspondence to study the orbit Ω of the vacuum vector 1 in B under the action of the group GL_∞:

$$\Omega = GL_\infty \cdot 1 . \tag{7.1}$$

The set Ω is an infinite-dimensional manifold, each point of which is, as we shall show, a solution of an infinite set of partial differential equations.

We are already familiar with a class of functions contained in Ω:

Proposition 7.1. The Schur polynomials $S_\lambda(x)$ ($\lambda \in Par$) are contained in Ω. *Proof.* By the correspondence between B and $F^{(0)}$, 1 is represented in $F^{(0)}$ by $\psi_0 = v_0 \wedge v_{-1} \wedge \ldots$ and $S_\lambda(x)$ by some $\psi_\lambda = v_{i_0} \wedge v_{i_{-1}} \wedge \ldots$, where $i_{-n} = -n$ for $n \geqslant$ some k. For $A \in GL_\infty$ defined by $Av_{-n} = v_{i_{-n}}$ for $0 \leqslant n \leqslant k-1$, and $Av_j = v_j$ for all other basis elements v_j, we have: $\psi_\lambda = R(A)\psi_0$. Hence for each $\lambda \in Par$, ψ_λ is in $GL_\infty \cdot \psi_0$, i.e. $S_\lambda \in \Omega$. $\blacksquare$

We shall use the symbol Ω to denote the orbit of the vacuum vector under GL_∞ interchangeably in B or in $F^{(0)}$.

7.2. Defining equations for Ω in $F^{(0)}$.

Proposition 7.2. If $\tau \in \Omega$, then τ is a solution of the equation

$$\sum_{j \in \mathbb{Z}} \hat{v}_j(\tau) \otimes \check{v}_j^*(\tau) = 0 . \tag{7.2}$$

Conversely, if $\tau \in F^{(0)}$, $\tau \neq 0$ and τ satisfies (7.2), then $\tau \in \Omega$.

69

Proof. $\hat{v}_j(\psi_0) = 0$ for $j < 0$ and $\check{v}_j^*(\psi_0)$ for $j > 0$ so that

$$\sum_{j \in \mathbb{Z}} \hat{v}_j(\psi_0) \otimes \check{v}_j^*(\psi_0) = 0 , \tag{7.3}$$

i.e. ψ_0 is a solution of (7.2). Any $\tau \in \Omega$ is of the form

$$\tau = R_0(A) \psi_0 , \tag{7.4}$$

where $A \in GL_\infty$. From their definitions (5.15) and (5.16) we easily see that the wedging and contracting operators have the following transformation properties under $R(A)$:

$$R_0(A)\hat{v} R_0(A)^{-1} = \hat{w} , \quad \text{where } w = Av , \tag{7.5a}$$

$$R_0(A)\check{f} R_0(A)_\cdot^{-1} = \check{g} , \quad \text{where } g = {}^tA^{-1}f , \tag{7.5b}$$

and ${}^tA^{-1}$ is the transpose of A^{-1}. We denote the matrix elements of A and A^{-1} in the basis $\{v_i \mid i \in \mathbb{Z} \}$ by a_{ij} and $\bar{a}_{ij}$ respectively, so that:

$$Av_j = \sum_i a_{ji}v_i ; \quad {}^tA^{-1}v_j^* = \sum_k \bar{a}_{kj}v_k^* ; \quad \sum_j \bar{a}_{kj}a_{ji} = \delta_{ki} . \tag{7.6}$$

If we apply $R(A)$ to (7.3) it will act on each component of the tensor product. Using (7.4) we get:

$$\sum_j R_0(A)\hat{v}_j R_0(A)^{-1} (\tau) \otimes R_0(A)\check{v}_j^* R_0(A)^{-1} (\tau) = 0 .$$

Using (7.5) and (7.6) this becomes:

$$\sum_{i,j,k} a_{ji}\hat{v}_i(\tau) \otimes \bar{a}_{kj} \check{v}_k^* (\tau) = 0$$

which can be rewritten as:

$$\sum_{i,k} \left(\sum_j \bar{a}_{kj}a_{ji} \right) \hat{v}_i (\tau) \otimes \check{v}_k^* (\tau) = 0$$

i.e., as (7.2).

Conversely, let $\tau \in F^{(0)}, \tau \neq 0$ and τ satisfy (7.2). We can write $\tau = \sum_{k=1}^{N} c_k \tau_k$, a linear combination with non-zero coefficients c_k of some semi-infinite monomials τ_k, such that τ_1 is a semi-infinite monomial of greatest (principal) degree; we may assume that $c_1 = 1$. If among the τ_i with $i > 1$ there exists a semi-infinite monomial, say τ_2, of the form

$$r_0(E_{ij})\tau_1 \quad \text{with} \quad i < j \ , \tag{7.7}$$

we can kill off the term $c_2\tau_2$ by replacing τ by $R_0(\exp - c_2 \, E_{ij})\tau$, which again satisfies (7.2) (as shown above). Repeating this procedure several (but a finite) number of times we arrive at an element of the form $\tau_1 + \varphi$, where none of the semi-infinite monomials appearing in φ is equal to τ_1 or is of the form (7.7). Since $\tau_1 + \varphi$ satisfies (7.2), it follows that $\varphi = 0$. Since, being a semi-infinite monomial, $\tau_1 \in \Omega$, we obtain that $\tau \in \Omega$. ∎

7.3. Differential equations for Ω in $\mathbb{C}[x_1, x_2, \ldots]$.

Consider the expression

$$X(u)\tau \otimes X^*(u)\tau \tag{7.8a}$$

where $X(u)$, $X^*(u)$ are the generating functions defined in (5.21a, b). Then (7.8a) can be rewritten as:

$$\sum_{i,j} u^{i-j} \, \hat{v}_i(\tau) \otimes \check{v}_j^*(\tau) \ , \tag{7.8b}$$

and it follows from Proposition 7.2 that $\tau \in \Omega$ if and only if the "constant term" (the term independent of u) in (7.8b) vanishes.

The isomorphism between $F^{(0)}$ and $\mathbb{C}[x_1, x_2, \ldots]$, which we discussed in Lecture 5, extends to an isomorphism between $F^{(0)} \otimes F^{(0)}$ and $\mathbb{C}[x_1', x_2', \ldots; x_1'', x_2'', \ldots]$, which is the polynomial ring in $x_1', x_2', \ldots, x_1'', x_2'', \ldots$. We can transform (7.8a) to the bosonic representation using the identification established in Proposition 5.1:

$$X(u) \ \rightarrow \ \Gamma(u) = uz \, \exp\left(\sum_{j \geqslant 1} u^j x_j' \right) \exp\left(- \sum_{j \geqslant 1} \frac{u^{-j}}{j} \frac{\partial}{\partial x_j'} \right) ,$$

$$X^*(u) \rightarrow \Gamma^*(u) = z^{-1} \exp\left(- \sum_{j \geqslant 1} u^j x_j'' \right) \exp\left(\sum_{j \geqslant 1} \frac{u^{-j}}{j} \frac{\partial}{\partial x_j''} \right) .$$

Thus (7.8a) becomes:

$$u \; \exp \left(\sum_{j \geqslant 1} u^j (x_j' - x_j'') \right) \exp \left(- \sum_{j \geqslant 1} \frac{u^{-j}}{j} \left(\frac{\partial}{\partial x_j'} - \frac{\partial}{\partial x_j''} \right) \right) \tau(x') \tau(x'') \; .$$

Defining new variables x, y by:

$$x' = x - y \, , \quad x'' = x + y \tag{7.9a}$$

so that

$$x' - x'' = - 2y \, , \quad \frac{\partial}{\partial x'} - \frac{\partial}{\partial x''} = - \frac{\partial}{\partial y} \, , \tag{7.9b}$$

we deduce from Proposition 7.2

Proposition 7.3. A nonzero element τ of $\mathbb{C}[x_1, x_2, \ldots]$ is contained in Ω if and only if the coefficient of u^0 vanishes in the expression:

$$u \; \exp \left(- \sum_{j \geqslant 1} 2u^j y_j \right) \exp \left(\sum_{j \geqslant 1} \frac{u^{-j}}{j} \frac{\partial}{\partial y_j} \right) \tau(x - y) \tau(x + y) \; . \tag{7.10}$$

∎

7.4. Hirota's bilinear equations.

Definition 7.1. Given a polynomial $P(x_1, x_2, \ldots)$ depending on a finite number of the $x_j \, (j = 1, 2, \ldots)$, and two functions f and g, we denote by $Pf \cdot g$ the expression

$$P(\frac{\partial}{\partial u_1}, \frac{\partial}{\partial u_2}, \ldots)(f(x_1 - u_1, x_2 - u_2, \ldots) g(x_1 + u_1, x_2 + u_2, \ldots)) \Big|_{u = 0} \; . \tag{7.11}$$

The equation $Pf \cdot g = 0$ is called a *Hirota bilinear equation.*

To illustrate this notation take $P = x$. Then

$$Pf \cdot g \; = \; \frac{\partial}{\partial u} (f(x - u) g(x + u)) \Big|_{u = 0} \; = \; - g(x) \frac{\partial f}{\partial x} + f(x) \frac{\partial g}{\partial x} \; .$$

Let $P = x^n$. Then from Leibniz's formula we obtain:

$$Pf \cdot g = \sum_{k=0}^{n} (-1)^k \binom{n}{k} \frac{\partial^k f}{\partial x^k} \frac{\partial^{n-k} g}{\partial x^{n-k}} \ .$$

Remark 7.1. Note that

$$Pf \cdot f \equiv 0 \quad \text{if and only if} \quad P(x) = -P(-x) \ .$$

We now expand the exponentials in (7.10) with the help of the generating functions (6.1) for elementary Schur polynomials:

$$u \left(\sum_{j \geq 0} u^j S_j(-2y) \right) \left(\sum_{j \geq 0} u^{-j} S_j(\tilde{\partial}_j) \right) \tau(x-y)\tau(x+y) \ , \qquad (7.12)$$

where

$$\tilde{\partial}_y = \left(\frac{\partial}{\partial y_1} , \frac{1}{2} \frac{\partial}{\partial y_2} , \frac{1}{3} \frac{\partial}{\partial y_3} , \dots \right) \ . \qquad (7.13)$$

Putting equal to zero the term in (7.12) which is independent of u, we get the system of equations:

$$\sum_{j \geq 0} S_j(-2y) S_{j+1}(\tilde{\partial}_y) \tau(x-y)\, \tau(x+y) = 0 \ . \qquad (7.14)$$

Now

$$S_{j+1}(\tilde{\partial}_y)\tau(x-y)\tau(x+y) = S_{j+1}(\tilde{\partial}_u)\tau(x-y-u)\tau(x+y+u)\Big|_{u=0}$$

$$= S_{j+1}(\tilde{\partial}_u)\exp(\sum_{s \geq 1} y_s \frac{\partial}{\partial u_s})\tau(x-u)\tau(x+u)\Big|_{u=0} \ ,$$

using Taylor's formula. This last expression can be written as

$$S_{j+1}(\tilde{x}) \exp\left(\sum_{s \geq 1} y_s x_s \right) \tau(x) \cdot \tau(x) \ ,$$

where

$$\tilde{x} = \left(x_1, \frac{1}{2}x_2, \frac{1}{3}x_3, \dots \right) \ .$$

We thus arrive at

Theorem 7.1. A nonzero polynomial τ is contained in Ω if and only if τ is a solution of the following system of Hirota bilinear equations:

$$\sum_{j=0}^{\infty} S_j(-2y) \, S_{j+1}(\widetilde{x}) \exp\left(\sum_{s \geqslant 1} y_s x_s \right) \tau(x) \cdot \tau(x) = 0 \ , \qquad (7.15)$$

where $y_1, y_2, \ldots$ are free parameters.

Proof follows now immediately from Proposition 7.3. ■

Equation (7.15) is due to Kashiwara and Miwa [1981].

7.5. The KP hierarchy.

If we expand (7.15) in a multiple Taylor series in the variables $y_1, y_2, \ldots$, then each coefficient of this series must vanish, giving us thereby a nonlinear partial differential equation. Let us take the simple case of determining the coefficient of y_r in this expansion. Expanding the exponential in (7.15) we see that y_r appears exactly once with coefficient x_r. In the expansion of the $S_j(-2y)$, y_r appears only in $S_r(-2y)$ with coefficient -2. Thus collecting the coefficient of y_r, we get the Hirota bilinear equation

$$(x_r x_1 - 2S_{r+1}(\widetilde{x})) \, \tau \cdot \tau = 0 \ . \qquad (7.16)$$

With the help of (6.3), we find that

$$x_r x_1 - 2S_{r+1}(\widetilde{x}) = -x_2 \quad \text{for} \ r = 1 \ ,$$

$$= -x_1^3/3 - 2x_3/3 \quad \text{for} \ r = 2 \ ,$$

$$= x_1 x_3/3 - x_4/2 - x_2^2/4 - x_1^4/12 - x_1^2 x_2/2 \quad \text{for} \ r = 3 \ .$$

From Remark 7.1 we can drop all odd polynomials. Hence $r = 1, 2$ give trivial equations, while for $r = 3$ the even terms give the Hirota equation

$$P\tau \cdot \tau = 0 \ , \ \text{where} \ P = x_1^4 + 3x_2^2 - 4x_1 x_3 \ . \qquad (7.17)$$

From (7.11) we can rewrite (7.17) as:

$$\left(\frac{\partial^4}{\partial u_1^4} + 3 \frac{\partial^2}{\partial u_2^2} - 4 \frac{\partial^2}{\partial u_1 \partial u_3}\right) \tau(x+u)\tau(x-u)\Big|_{u=0} = 0 \ .$$

Putting $x_1 = x, x_2 = y, x_3 = t$ and introducing a new function

$$u(x, y, t) = 2 \frac{\partial^2}{\partial x^2} (\log \tau) \ ,$$

we find after a calculation that (7.17) becomes the Kadomtzev-Petviashvili (KP) equation:

$$\frac{3}{4} \frac{\partial^2 u}{\partial y^2} = \frac{\partial}{\partial x}\left(\frac{\partial u}{\partial t} - \frac{3}{2} u \frac{\partial u}{\partial x} - \frac{1}{4} \frac{\partial^3 u}{\partial x^3}\right) \ . \tag{7.18}$$

Note that the term in brackets on the right-hand side of (7.18) is the KdV equation. Hence, if u is independent of y, the KP equation reduces essentially to the KdV equation.

As an immediate consequence of Proposition 7.1 and Theorem 7.1 we have:

Corollary 7.1. The following functions are rational solutions of the KP equation:

$$2 \frac{\partial^2}{\partial x^2} (\log S_\lambda(x, y, t, c_4, c_5, \ldots)) \ ,$$

where $c_4, c_5, \ldots$ are arbitrary constants. ∎

Remark 7.2. The family of nonlinear equations (7.15), of which the first is the KP equation, is known as the *KP hierarchy*. The idea that the solutions of the KP hierarchy are parametrised by an infinite dimensional homogeneous space is due to Sato [1981], and it was developed by Date-Jimbo-Kashiwara-Miwa [1981]–[1983]. The KP hierarchy appears as the compatibility condition for the system of differential equations

$$\frac{\partial}{\partial x_n} \varphi(x, u) = A_+^n(x, \partial)\varphi(x, u)$$

where A_+^n denotes the differential part (i.e. terms containing non-negative powers of $\partial \equiv \partial/\partial x_1$) of the n-th power of the formal pseudodifferential operator

$$A = \partial + a_1(x)\partial^{-1} + a_2(x)\partial^{-2} + \ldots .$$

Here $x = (x_1, x_2, \ldots)$, ∂^{-1} is the formal inverse of $\partial/\partial x_1$, and $\varphi(x, u) = 1 + \varphi_1(x)u^{-1} + \varphi_2(x)u^{-2} + \ldots .$ The compatibility conditions are

$$\frac{\partial A_+^m}{\partial x_n} - \frac{\partial A_+^n}{\partial x_m} = [A_+^n, A_+^m] .$$

Equating the coefficients of powers of ∂ in the above equation gives the KP hierarchy of PDE's on the functions a_i. We can also rewrite the compatibility conditions in the Lax form

$$\frac{\partial A}{\partial x_n} = [A_+^n, A] .$$

The function $\varphi(x, u)$ and the operator A are related to $\tau \in \Omega$ by $\varphi(x, u) = \tau(x)^{-1} \left(\exp \sum_{j \geqslant 1} u^j x_j \right) \left(\exp - \sum_{j \geqslant 1} \frac{u^{-j}}{j} \frac{\partial}{\partial x_j} \right) \tau(x)$, and $L = P\partial P^{-1}$, where $P = 1 + \varphi_1(x)\partial^{-1} + \varphi_2(x)\partial^{-2} + \ldots$ (see Kashiwara-Miwa [1981] for details).

Remark 7.3. Denote by UGM (universal Grassmann manifold) the set of all subspaces U of $V = \sum_{j \in \mathbb{Z}} \mathbb{C}v_j$ such that U contains $\sum_{j \leqslant -k} \mathbb{C}v_j$ for $k \gg 0$ as a subspace of codimension k. Note that $\Omega = \{u_0 \frown u_{-1} \frown \ldots | u_{-j} = v_{-j}$ for $j \gg 0 \}$, so that we can define a bijective map

$$f: \mathbb{P}\Omega \xrightarrow{\sim} UGM$$

by $f(u_0 \frown u_{-1} \frown \ldots) = \sum_j \mathbb{C}u_j$ (here $\mathbb{P}\Omega$ stands for projectivisation of Ω).

Thus (due to Theorem 7.1) the set of non-zero polynomial solutions (considered up to a constant factor) of the KP hierarchy is parametrized by the UGM. This is the fundamental observation of Sato [1981].

7.6. *N*-soliton solutions.

The Lie algebra $g\ell_\infty$ has a representation in $\mathbb{C}[x_1, x_2, \ldots]$ defined by the vertex operator $\Gamma(u, v)$. Exponentiating an element of $g\ell_\infty$ gives us an element of GL_∞. The following proposition shows that $\exp(a\Gamma(u, v))$ may be thought of as $1 + a\Gamma(u, v)$:

Proposition 7.4. $\Gamma(u, v)^2 \tau = 0$ for 'good' formal power series τ.

Proof. By Taylor's formula,

$$\exp\left(\sum_{i \geqslant 1} \lambda_i \frac{\partial}{\partial x_i}\right) \tau(x_1, x_2 \ldots) = \tau(x_1 + \lambda_1, x_2 + \lambda_2, \ldots) \; .$$

Hence

$$\Gamma(u, v)\tau(\ldots, x_j, \ldots) = (\exp \sum_{j \geqslant 1} (u^j - v^j)x_j)\tau(\ldots, x_j - \frac{u^{-j} - v^{-j}}{j}, \ldots) \; .$$

Using the well-known commutation relation

$$e^A e^B = e^B e^A e^C \; ,$$

which holds if

$$[A, B] = cI \; ,$$

and the expansion

$$\log(1 - z) = -\sum_{j \geqslant 1} \frac{z^j}{j} \quad (|z| < 1)$$

we find, under the assumption $|u|, |v| < \min(|u'|, |v'|)$:

$$\Gamma(u', v') \Gamma(u, v)\tau(\ldots, x_j, \ldots) = \frac{(u' - u)(v' - v)}{(v' - u)(u' - v)}$$

$$\times \exp(\sum_{j \geqslant 1} (u^j - v^j + u'^j - v'^j)x_j) \, \tau(\ldots, x_j - \frac{u^{-j} - v^{-j} + u'^{-j} - v'^{-j}}{j}, \ldots) \; .$$

$$(7.19)$$

The expression (7.19) is valid for all $u \neq v'$, $v \neq u'$ by analytic continuation. Taking the limit $u' \to u$, $v' \to v$ in (7.19), we get $\Gamma(u, v)^2 \, \tau = 0$. ∎

Corollary 7.2. The function

$$\tau_{N;a;u,v}(x) = (1 + a_1\Gamma(u_1, v_1)) \ldots (1 + a_N\Gamma(u_N, v_N)) \cdot 1 \quad (7.20)$$

is a solution of the KP hierarchy. (It is known as the τ-function of an *N-soliton solution.*)

Proof. The KP hierarchy can be written symbolically as follows:

$$S(\tau \otimes \tau) = 0 \, ,$$

where $S = \sum\limits_{j \in \mathbb{Z}} \hat{v}_j \otimes \check{v}_j^*$ is an operator on $F \otimes F$ commuting with the diagonal action of GL_∞ (see the proof of Proposition 7.2). Since $\Gamma(u, v)$ lies in the completion of $g\ell_\infty$ (Proposition 5.2), we have:

$$\Gamma(u, v)S(\tau \otimes \tau) = S(\Gamma(u, v)\tau \otimes \tau + \tau \otimes \Gamma(u, v)\tau) \, . \quad (7.21)$$

Since

$$2\Gamma(u, v)\tau \otimes \Gamma(u, v)\tau = \Gamma(u, v)^2 \, (\tau \otimes \tau) - \Gamma(u, v)^2 \, \tau \otimes \tau - \tau \otimes \Gamma(u, v)^2 \, \tau = 0$$

by Proposition 7.4, we deduce from (7.21) that

$$S((1 + \Gamma(u, v)\tau) \otimes (1 + \Gamma(u, v)\tau) = S(\tau \otimes \tau) + \Gamma(u, v)S(\tau \otimes \tau) \, ,$$

This shows that if τ is a solution of the KP hierarchy, then $1 + a\Gamma(u, v)\tau$ is one as well. Since $\tau = 1$ is a solution, the proof is completed. ∎

The 1-soliton solution of the KP equation (7.18) is given by

$$u(x, y, t) = 2 \, \frac{\partial^2 (\log \tau_{1;a;u,v}(x))}{\partial x^2} \, ,$$

where

$$\tau_{1;a;u,v} = (1 + a\Gamma(u, v)) \cdot 1$$

$$= 1 + \exp((u - v)x + (u^2 - v^2)y + (u^3 - v^3)t + c) \, ,$$

and c is a constant. Thus, we obtain that

$$u(x, y, t) = \frac{(u - v)^2}{2} \cdot \frac{1}{[\cosh \frac{1}{2}((u - v)x + (u^2 - v^2)y + (u^3 - v^3)t + c)]^2}$$

is a 1-soliton solution of the KP equation.

One easily expands (7.21) using (7.19) to write an explicit formula for the N-soliton solutions (see Date-Jimbo-Kashiwara-Miwa [1981]). We have from (7.19):

$$\Gamma(u_1, v_1) \ldots \Gamma(u_N, v_N) \cdot 1 = \prod_{1 \leqslant i < j \leqslant N} \frac{(u_j - u_i)(v_j - v_i)}{(u_j - v_i)(v_j - u_i)}$$

$$\times \exp\left(\sum_{j=1}^{\infty} \sum_{k=1}^{N} (u_k^j - v_k^j)x_j \right).$$

Hence we obtain:

$$\tau_{N;a;u,v}(x) = \sum_{\substack{0 \leqslant r \leqslant N \\ 1 \leqslant j_1 < j_2 < \ldots < j_r \leqslant N}} \prod_{s=1}^{r} a_{j_s} \prod_{1 \leqslant \lambda < \mu \leqslant r} \frac{(u_{j_\lambda} - u_{j_\mu})(v_{j_\lambda} - v_{j_\mu})}{(u_{j_\lambda} - v_{j_\mu})(v_{j_\lambda} - u_{j_\mu})}$$

$$\times \exp \sum_{k \geqslant 1} \sum_{m=1}^{r} (u_{j_m}^k - v_{j_m}^k)x_k,$$

so that $2(\partial^2/\partial x^2)(\log \tau_{N;a;u,v}(x, y, t, c_4, c_5, \ldots))$ is the N-soliton solution of the KP equation.

LECTURE 8

8.1. Degenerate representations and the determinant $\det_n(c, h)$ of the contravariant form.

We saw in Lecture 3 that to every pair of real numbers (c, h) there corresponds the Verma representation $M(c, h)$, which carries a contravariant Hermitian form $\langle \cdot \mid \cdot \rangle$ and is such that any other highest weight representation is a quotient of $M(c, h)$. Quotienting $M(c, h)$ by its unique maximal proper subrepresentation $J(c, h)$ ($= \text{Ker} \langle \cdot \mid \cdot \rangle$) we get the unique irreducible representation $V(c, h)$ with highest weight (c, h).

It is a mathematically interesting question to determine when $M(c, h) = V(c, h)$, i.e. when $M(c, h)$ is irreducible. This problem was solved by Kac [1978]. It is clear that the answer can only depend on the highest weight (c, h). We shall see that generically $M(c, h) = V(c, h)$. If $V(c, h) \neq M(c, h)$ we shall say that $V(c, h)$ is a *degenerate representation* of Vir. In a remarkable recent development, the degenerate representations of Vir have acquired a special significance in the study of the critical behaviour of two dimensional statistical mechanical systems (Belavin-Polyakov-Zamolodchikov [1984a, b]). The classification of the degenerate representations of Vir is, therefore, of interest both in mathematics and physics.

From Proposition 3.4(c) we observe that, for $V(c, h)$ to be degenerate, the contravariant Hermitian form $\langle \cdot \mid \cdot \rangle$ on $M(c, h)$ must have a nontrivial kernel. Vectors of the form (3.12c) form a linearly independent set of vectors which span $M(c, h)$ and in this basis the matrix of the contravariant form $\langle \cdot \mid \cdot \rangle$ is

$$(\langle d_{-i_t} \cdots d_{-i_1} (v) \mid d_{-j_s} \cdots d_{-j_1} (v) \rangle) , \qquad (8.1a)$$

where

$$1 \leqslant i_1 \leqslant \ldots \leqslant i_t , \quad 1 \leqslant j_1 \leqslant \ldots \leqslant j_s , \qquad (8.1b)$$

and v is the highest weight vector. However, from Proposition 3.4(b) we know that $M(c, h)$ is a direct sum of finite dimensional eigenspaces of d_0 which are mutually orthogonal with respect to $\langle \cdot \mid \cdot \rangle$. Hence the matrix of $\langle \cdot \mid \cdot \rangle$ is a direct sum of finite-dimensional matrices in each eigenspace $M(c, h)_{h+n}$ of d_0 with eigenvalue $h + n$, $n \in \mathbb{Z}_+$. The restriction of $\langle \cdot \mid \cdot \rangle$ to $M(c, h)_{h+n}$, the *n-th level*, is the $p(n) \times p(n)$ matrix defined by (8.1a, b) with the additional condition that

$$\sum_k i_k = \sum_k j_k = n \ . \tag{8.1c}$$

Note that the entries of this matrix are polynomials in c and h.

The condition for the representation $V(c, h)$ to be unitary is that the matrix (8.1a, b, c) should be positive semi-definite for each $n \in \mathbb{Z}_+$. It turns out that this gives us an effective means of determining for which highest weights $V(c, h)$ is unitary. We have an even simpler criterion for degeneracy: a necessary and sufficient condition for the degeneracy of $V(c, h)$ is that for some $n \in \mathbb{Z}_+$ the determinant of the matrix defined by (8.1a, b, c) should vanish. We shall denote the determinant of the matrix defined by (8.1a, b, c) by $\det_n(c, h)$.

The first few values of $\det_n(c, h)$ are easily found:

$$\det_0(c, h) = \langle v \mid v \rangle = 1,$$

$$\det_1(c, h) = \langle d_{-1} v \mid d_{-1} v \rangle = 2h \ ,$$

$$\det_2(c, h) = \begin{vmatrix} \langle d_{-2} v \mid d_{-2} v \rangle & \langle d_{-2} v \mid d_{-1}^2 v \rangle \\ \langle d_{-1}^2 v \mid d_{-2} v \rangle & \langle d_{-1}^2 v \mid d_{-1}^2 v \rangle \end{vmatrix}$$

$$= \begin{vmatrix} 4h + c/2 & 6h \\ 6h & 8h^2 + 4h \end{vmatrix}$$

$$= 2h(16h^2 + 2hc - 10h + c) \ . \tag{8.2}$$

A necessary condition for the representation $V(c, h)$ to be unitary is that $\det_n(c, h) \geqslant 0$ for $n = 0, 1, 2, \ldots$. Thus from $n = 1$ we find that $h \geqslant 0$ is a necessary condition, as we already noted in Proposition 3.5, where we saw that $c \geqslant 0$ is also a necessary condition. The $n = 2$ case gives us more precise information: rewriting $16h^2 + 2hc - 10h + c$ as $(4h - 1)^2 + (2h + 1)(c - 1)$ we observe that the region of the c-h plane defined by

$$0 \leqslant c < 1 - (4h - 1)^2/(2h + 1), \quad h \geqslant 0$$

is a region of non-unitarity (see Fig. 8.1). Thus by studying the sign of $\det_n(c, h)$ we can determine regions of the c-h plane where $M(c, h)$ is not unitary. To proceed further we need a general formula for $\det_n(c, h)$.

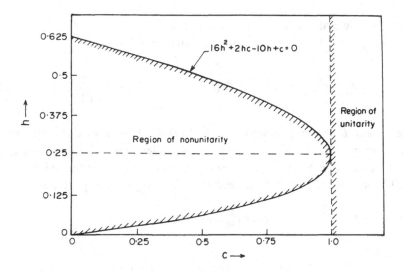

Figure 8.1

8.2. The determinant $\det_n(c, h)$ as a polynomial in h.

As the first step in finding a general expression for $\det_n(c, h)$, we shall fix the value of c and consider $\det_n(c, h)$ as a polynomial in h; we shall determine the degree of this polynomial.

Now, h comes from the action of d_0 on the highest weight vector v, and in the Virasoro algebra the commutator $[d_i, d_j]$ gives rise to d_0 only for $i + j = 0$. Examining the matrix (8.1a, b, c) we observe that in each column or row the term giving rise to the maximum number of such commutators is the term lying on the principal diagonal. Thus the leading term in h of $\det_n(c, h)$ is contained in the expression

$$\prod_{\substack{1 \leqslant i_1 \leqslant \ldots \leqslant i_s \\ \Sigma i_k = n}} \langle d_{-i_s} \ldots d_{-i_1}(v) \, | \, d_{-i_s} \ldots d_{-i_1}(v) \rangle . \qquad (8.3)$$

Lemma 8.1.

$$\langle d_{-n}^k(v) \mid d_{-n}^k(v) \rangle = k! n^k (2h + \frac{n^2 - 1}{12} c)(2h + \frac{n^2 - 1}{12} c + n) \ldots$$

$$\times \ (2h + \frac{n^2 - 1}{12} c + n(k - 1)) \tag{8.4}$$

where $n, k \in \mathbb{N}$ and v is the highest weight vector.

Proof is by induction on k, using the commutation relation

$$[d_n, d_{-n}^k] = nk d_{-n}^{k-1} (n(k - 1) + 2d_0 + \frac{n^2 - 1}{12} c) \tag{8.5}$$
∎

Let A and B be two polynomials in h. In the following we shall write $A \sim B$ if the term with the highest power of h in A is identical to the term having the highest power of h in B. This is clearly an equivalence relation and if $A \sim C$, $B \sim D$ then $AB \sim CD$. We shall need the following corollary of Lemma 8.1:

Corollary 8.1. $\langle d_{-n}^k(v) \mid d_{-n}^k(v) \rangle \sim k!(2nh)^k$. $\tag{8.6}$
∎

Lemma 8.2.

$$\langle d_{-i_s}^{j_s} \ldots d_{-i_1}^{j_1}(v) \mid d_{-i_s}^{j_s} \ldots d_{-i_1}^{j_1}(v) \rangle$$

$$\sim \langle d_{-i_s}^{j_s}(v) \mid d_{-i_s}^{j_s}(v) \rangle \ldots \langle d_{-i_1}^{j_1}(v) \mid d_{-i_1}^{j_1}(v) \rangle ,$$

where $i_1, \ldots, i_s, j_1, \ldots, j_s \in \mathbb{N}$ and $i_1 \neq i_2 \neq \ldots \neq i_s$.

Proof is by induction on $\sum_k j_k$. ∎

If we now apply Lemma 8.2 to (8.3) we get:

Lemma 8.3. $\det_n(c, h) \sim \prod_{\substack{r, s \in \mathbb{N} \\ 1 \leqslant rs \leqslant n}} \langle d_{-r}^s(v) \mid d_{-r}^s(v) \rangle^{m(r, s)}$

where $m(r, s)$ is the number of partitions of n in which r appears exactly s times. ∎

Proposition 8.1. $\det_n(c, h)$ is, for fixed c, a polynomial in h of degree

$$\prod_{\substack{r, s \in \mathbb{N} \\ 1 \leqslant rs \leqslant n}} p(n - rs) , \tag{8.7}$$

and the coefficient K of the highest power of h is given by

$$K = \prod_{\substack{r, s \in \mathbb{N} \\ 1 \leqslant rs \leqslant n}} ((2r)^s s!)^{m(r, s)} , \tag{8.8a}$$

where $$m(r, s) = p(n - rs) - p(n - r(s + 1)) . \tag{8.8b}$$

Proof. From Lemma 8.3 and Corollary 8.1 we get the degree of $\det_n(c, h)$ in h to be

$$\sum_{\substack{r, s \in \mathbb{N} \\ 1 \leqslant rs \leqslant n}} s \, m(r, s)$$

with the leading coefficient (8.8a). We thus only have to compute $m(r, s)$, which is the number of partitions of n in which r appears exactly s times. The number of partitions of n in which r appears at least s times in clearly $p(n - rs)$. We subtract from this the number of partitions of n in which r appears at least $s + 1$ times, *viz.* $p(n - r(s + 1))$ to get (8.8b). Then,

$$\sum_{\substack{r, s \in \mathbb{N} \\ 1 \leqslant rs \leqslant n}} s \, m(r, s) = \sum_{1 \leqslant r \leqslant n} \sum_{s=1}^{[n/r]} s(p(n - rs) - p(n - r(s + 1)))$$

$$= \sum_{1 \leqslant r \leqslant n} \sum_{s=1}^{[n/r]} p(n - rs) = \sum_{\substack{r, s \in \mathbb{N} \\ 1 \leqslant rs \leqslant n}} p(n - rs).$$

(The symbol $[n/r]$ means the largest integer not exceeding n/r). ∎

8.3. The Kac determinant formula.

We shall require the following simple lemma in linear algebra:

Lemma 8.4. Let $A(t)$ be a family of linear operators acting in an n-dimensional vector space V and suppose that $A(t)$ is a polynomial function of t.

If $A(0)$ has a null space of dimension k, then $\det A(t)$ is divisible by t^k.

Proof. We choose the basis $e_1, \ldots, e_k, e_{k+1}, \ldots, e_n$ in V, such that $e_1, \ldots, e_k$ span the null space of $A(0)$, so that $A(0)e_i = 0$ for $1 \leqslant i \leqslant k$. Then, the first k rows of the matrix of $A(t)$ are divisible by t, proving the lemma. ∎

Lemma 8.5. Consider $\det_n(c, h)$ as a polynomial in h for fixed c. Suppose $\det_n(c, h)$ has a zero at $h = h_0$. Then $\det_n(c, h)$ is divisible by

$$(h - h_0)^{p(n-k)}$$

where k is the smallest positive integer $(1 \leqslant k \leqslant n)$ for which $\det_k(c, h)$ vanishes at $h = h_0$.

Proof. If $\det_n(c, h)$ vanishes at $h = h_0$, then by Proposition 3.4 the matrix of the contravariant form $\langle \cdot \mid \cdot \rangle$ on $M(c, h_0)$ has a nonzero kernel when restricted to the n-th level. Thus $M(c, h_0)$ has a nonzero maximal proper subrepresentation $J(c, h_0)$ with a nonzero component $J_n(c, h_0)$ in the n-th level. Let $k \in \mathbb{Z}_+$ be the smallest number such that $J_k(c, h_0) \neq 0$. Picking a non-zero u in $J_k(c, h_0)$, we have:

$$d_0(u) = (h_0 + k)u \quad \text{and} \quad d_n(u) = 0 \quad \text{for } n > 0 .$$

Hence u is a singular vector. The application of the universal enveloping algebra $U(Vir)$ to u generates a subrepresentation of $M(c, h_0)$ which is contained in $J(c, h_0)$. The component of this subrepresentation in the n-th level is the linear span of vectors of the form

$$d_{-i_s} \cdots d_{-i_1}(u) \quad (0 < i_1 \leqslant \ldots \leqslant i_s, \sum_s i_s = n - k) . \tag{8.9}$$

All such vectors are linearly independent. This follows from the standard fact that a singular vector of a Verma representation generates again a Verma representation. (The latter fact follows from the absence of zero divisors in $U(Vir)$.) The vectors (8.9) thus span a subspace of $J_n(c, h_0)$ of dimension $p(n-k)$. Hence the matrix of the contravariant form restricted to level n has a kernel of dimension at least $p(n-k)$. By Lemma 8.4 it follows that $\det_n(c, h)$ is divisible by $(h - h_0)^{p(n-k)}$. Since u lies in level k, $J(c, h_0)$ has a nontrivial component in level k and $\det_k(c, h)$ must vanish at $h = h_0$. By the definition of u, k is the minimum value for which this happens. ∎

We also require the following lemma, the proof of which we defer to Lecture 12.

Lemma 8.6. Considered as a polynomial in h, $\det_n(c, h)$ has a zero at $h = h_{r,s}(c)$, where

$$h_{r,s}(c) = \frac{1}{48} \left[(13 - c)(r^2 + s^2) + \sqrt{(c-1)(c-25)} \, (r^2 - s^2) \right.$$

$$\left. - 24rs - 2 + 2c \right] , \tag{8.10}$$

for each pair (r, s) of positive integers such that $1 \leqslant rs \leqslant n$.

Corollary 8.2. $\det_n(c, h)$ is divisible by

$$\Phi = \prod_{\substack{r, s \in \mathbb{N} \\ 1 \leqslant rs \leqslant n}} (h - h_{r,s}(c))^{p(n-rs)} \tag{8.11}$$

Proof. It follows from Lemma 8.6 that $\det_k(c, h)$ has a zero at $h = h_{r,s}(c)$ for $rs \leqslant k \leqslant n$. The corollary now follows from Lemma 8.5. ∎

Theorem 8.1 (Kac [1978]).

$$\det_n(c, h) = K \prod_{\substack{r, s \in \mathbb{N} \\ 1 \leqslant rs \leqslant n}} (h - h_{r,s}(c))^{p(n-rs)} \tag{8.12}$$

where $h_{r,s}(c)$ is given by (8.10) and K is the positive constant given by (8.8a, b) (which depends only on n).

Alternatively, let

$$\varphi_{r,r} = h - h_{r,r} = h + (r^2 - 1)(c - 1)/24 , \tag{8.13a}$$

and for $r \neq s$ let

$$\varphi_{r,s} = (h - h_{r,s})(h - h_{s,r}) . \tag{8.13b}$$

Then

$$\det_n(c, h) = K \prod_{\substack{r, s \in \mathbb{N} \\ s \leqslant r \\ 1 \leqslant rs \leqslant n}} \varphi_{r,s}^{p(n-rs)} . \tag{8.14}$$

Proof. From Corollary 8.2 we know that $\det_n(c, h)$ is divisible by Φ given by (8.11). Moreover the degree of Φ in h agrees with the degree of $\det_n(c, h)$ in h given by (8.7). Hence $\det_n(c, h)$ and Φ can only differ by an overall constant which could depend on c. This constant is the coefficient of the highest power of h which has, however, already been computed in (8.8a, b) and is independent of c. ∎

Corollary 8.3. If $\varphi_{r, s}(c, h) = 0$ and $\varphi_{r', s'}(c, h) \neq 0$ for $r's' < rs$, then $M(c, h)$ has a singular vector of level rs.

Proof. An immediate consequence of the proof of Lemma 8.5 and Theorem 8.1. ∎

Equation (8.14) is very convenient for computations. For $n = 2$, $\det_2(c, h) = 32 \, \varphi_{2, 1} \, \varphi_{1, 1}$ which agrees with (8.2).

8.4. Some consequences of the determinant formula for unitarity and degeneracy.

Proposition 8.2. (a) The irreducible highest weight representation $V(c, h)$ of the Virasoro algebra is unitary for $c \geqslant 1$ and $h \geqslant 0$.
(b) $V(c, h) = M(c, h)$ for $c > 1, h > 0$.

Proof. To prove (b), it suffices to show that for each $n \in \mathbb{Z}_+$, $\det_n(c, h) > 0$ for $c > 1, h > 0$. Now for $1 \leqslant r \leqslant n$, we have $\varphi_{r, r} = h + (r^2 - 1)(c - 1)/24 > 0$ for $c > 1, h > 0$. For $r \neq s$, $\varphi_{r, s}$ can be rewritten as

$$\varphi_{r, s} = \left(h - \frac{(r - s)^2}{4} \right)^2 + \frac{h}{24} (r^2 + s^2 - 2)(c - 1)$$

$$+ \frac{1}{576} (r^2 - 1)(s^2 - 1)(c - 1)^2 + \frac{1}{48} (c - 1)(r - s)^2 (rs + 1) . \tag{8.15}$$

Thus, $\varphi_{r, s} > 0$ for $1 \leqslant rs \leqslant n, c > 1, h > 0$. Hence from (8.14), $\det_n(c, h) > 0$ for $c > 1, h > 0$, proving (b).

The positivity of the determinants $\det_n(c, h)$ $(n \in \mathbb{Z}_+)$ in the region $c > 1$, $h > 0$ implies that if we can show that the contravariant form is positive definite at a single point in the region, then it is positive definite throughout the region

$c > 1$, $h > 0$ and hence positive semidefinite on $M(c, h)$ (so that $V(c, h)$ is unitary) throughout the region $c \geqslant 1$, $h \geqslant 0$. We have seen, however, in Lecture 3 that we have a manifestly unitary representation of Vir in terms of bosonic oscillators for $c = 1, 2, 3, \ldots$ and $h \geqslant 0$, proving (a). ∎

Proposition 8.3.   (a) $V(1, h) = M(1, h)$ if and only if $h \neq m^2/4 \ (m \in \mathbb{Z})$.

                 (b) $V(0, h) = M(0, h)$ if and only if $h \neq (m^2 - 1)/24 \ (m \in \mathbb{Z})$.

Proof.   For $c = 1$ formula (8.12) turns into:

$$\det{}_n(1, h) = K \prod_{\substack{r, s \in \mathbb{N} \\ 1 \leqslant rs \leqslant n}} \left(h - \frac{(r - s)^2}{4} \right)^{p(n - rs)} \tag{8.16a}$$

so that $\det_n(1, h) \neq 0$ for all $n \in \mathbb{Z}_+$ if and only if $h \neq m^2/4, m \in \mathbb{Z}$.

For $c = 0$ formula (8.12) turns into:

$$\det{}_n(0, h) = K \prod_{\substack{r, s \in \mathbb{N} \\ 1 \leqslant rs \leqslant n}} \left(h - \frac{(3r - 2s)^2 - 1}{24} \right)^{p(n - rs)} \tag{8.16b}$$

so that $\det_n(0, h) \neq 0$ for all $n \in \mathbb{Z}_+$ if and only if $h \neq (m^2 - 1)/24 \ (m \in \mathbb{Z})$. ∎

We have shown in Proposition 8.2 (following Kac [1982]) that $V(c, h)$ is unitary in the sector $c \geqslant 1$, $h \geqslant 0$ of the $c-h$ plane. We know from Proposition 3.5 that a necessary condition for $V(c, h)$ to be unitary is that $c \geqslant 0$, $h \geqslant 0$. This, therefore, leave the region $0 \leqslant c < 1$, $h \geqslant 0$ to be discussed.

A simple argument due to Gomes (see Goddard-Olive [1986]) shows that for $c = 0$ the only unitary highest weight representation of Vir is the trivial representation in which each d_n is represented by 0. For if the matrix of the contravariant form at level $2N$ is to be positive definite, we require in particular that the matrix

$$\begin{bmatrix} \langle d_{-2N}(v) \mid d_{-2N}(v) \rangle & \langle d^2_{-N}(v) \mid d_{-2N}(v) \rangle \\[2ex] \langle d_{-2N}(v) \mid d^2_{-N}(v) \rangle & \langle d^2_{-N}(v) \mid d^2_{-N}(v) \rangle \end{bmatrix}$$

be positive definite. Evaluating the determinant for $c = 0$ we obtain

$$4 N^3 h^2 (8h - 5N) ,$$

which is negative for large N unless $h = 0$. If $h = 0$, then $V(0, 0)$ is the 1-dimensional trivial representation.

In the region $0 \leqslant c < 1$ it is convenient to use the following parametrization of c:

$$c(m) = 1 - \frac{6}{(m + 2)(m + 3)} . \qquad (8.17)$$

The region $0 \leqslant c < 1$ corresponds to $m \geqslant 0$. This parametrization has the effect of rationalizing the expression (8.10) for $h_{r,s}$:

$$h_{r,s}(m) = \frac{((m + 3)r - (m + 2)s)^2 - 1}{4(m + 2)(m + 3)} . \qquad (8.18)$$

Now from Fig. 8.1 we know that not all points in the region $0 \leqslant c < 1, h > 0$ can correspond to unitary highest weight representations of *Vir*. Considering that the $n = 2$ case alone eliminates a large region, we might well suspect that the infinite number of nonlinear constraints coming from higher levels rule out the entire region. We know, however, from Lecture 3 that $c = \frac{1}{2}$ and $h = 0$, 1/16 or $\frac{1}{2}$ give rise to nontrivial unitary representations. By a detailed analysis of the Kac determinant formula, Friedan-Qiu-Shenker [1985], [1986] have shown that the only *possible* places of unitarity of $V(c, h)$ in the region $0 \leqslant c < 1$, $h \geqslant 0$, are the discrete set of points:

$$(c(m), h_{r,s}(m)), \quad \text{where } m, r, s \in \mathbb{Z}_+ \text{ and } 1 \leqslant s \leqslant r \leqslant m + 1. \quad (8.19)$$

For $m = 0$ we get $c = h = 0$, which is the trivial representation. For $m = 1$ we get $c = \frac{1}{2}$ and $h = 0, \frac{1}{2}, 1/16$ in agreement with our construction in Lecture 3.

We shall show in Lecture 12 that the representations $V(c, h)$ of *Vir* are indeed unitary for every pair (c, h) belonging to the "discrete series" (8.19).

Remark 8.1. Belavin-Polyakov-Zamolodchikov [1984b] pointed out that the series (8.19) correspond to the most important 2-dimensional statistical mechanical models. Then, Friedan-Qiu-Shenker [1985] interpreted this set from the point of view of unitarity property.

Remark 8.2. Another interesting region of irreducibility of the $M(c, h)$ is: $c \leqslant 1, 24h < c - 1$. It is easy to see that in this region one has:

$$\sum_{n \in \mathbb{Z}_+} s_n q^n = \prod_{n \geqslant 1} (1 + q^n)^{-1} \ ,$$

where s_n denotes the signature of $\langle \cdot \mid \cdot \rangle$ on the n-th level. This is shown by replacing in (3.24), λ and μ by $i\lambda$ and $i\mu$ and taking $\omega(a_n) = -a_{-n}$. The calculation of the series $\sum s_n q^n$ for arbitrary $M(c, h)$ is a very interesting open problem.

LECTURE 9

9.1. Representations of loop algebras in $\overline{a}_\infty$.

In Lecture 4 we realized the Lie algebra d as a subalgebra of $\overline{a}_\infty$, and, using this, we then constructed highest weight representations in $F^{(m)}$ of its central extension. We shall follow this procedure now for loop algebras.

Definition 9.1. Let $g\ell_n$ denote the Lie algebra of all $n \times n$ matrices with complex entries acting in $\mathbb{C}^n$ and let $\mathbb{C}[t, t^{-1}]$ denote the ring of Laurent polynomials (i.e. polynomials in t and t^{-1}). We define the *loop algebra* $\widetilde{g\ell}_n$ as $g\ell_n(\mathbb{C}[t, t^{-1}])$ i.e. as the complex Lie algebra of $n \times n$ matrices with Laurent polynomials as entries.

Remark 9.1. We can view $\widetilde{g\ell}_n$ as the Lie algebra of maps from the unit circle S^1 to the Lie algebra $g\ell_n$, with finite Fourier series, and the Lie bracket defined pointwise. This accounts for the name 'loop algebra'.

An element of $\widetilde{g\ell}_n$ has the form

$$a(t) = \sum_k t^k a_k \quad (a_k \in g\ell_n) , \qquad (9.1)$$

where k runs over a finite subset of $\mathbb{Z}$. Since the matrices $e_{ij}(1 \leqslant i, j \leqslant n)$, which have 1 as the (i, j) entry and 0 elsewhere, form a basis of $g\ell_n$, it is clear that the matrices

$$e_{ij}(k) \equiv t^k e_{ij} \quad (1 \leqslant i, j \leqslant n \text{ and } k \in \mathbb{Z}) \qquad (9.2)$$

form a basis of $\widetilde{g\ell}_n$. The elements of $\widetilde{g\ell}_n$ form an associative algebra with multiplication defined on the basis elements by

$$e_{ij}(k)e_{mn}(\ell) = t^{k+\ell} e_{ij} e_{mn} = \delta_{jm} e_{in}(k + \ell) . \qquad (9.3)$$

93

The Lie bracket on $\widetilde{g\ell}_n$ is the commutator of the associative multiplication defined by (9.3):

$$[\, e_{ij}(k),\, e_{mn}(\ell)\,] \;=\; \delta_{jm}\, e_{in}(k + \ell) - \delta_{ni}\, e_{mj}(k + \ell) \;. \qquad (9.4)$$

The vector space in which $g\ell_n$ acts is $\mathbb{C}^n$, which has a standard basis $u_1, \dots, u_n$ of $n \times 1$ column vectors in which $u_k\,(1 \leqslant k \leqslant n)$ has 1 in the k-th row and 0 elsewhere. The loop algebra $\widetilde{g\ell}_n$ acts in $\mathbb{C}[t, t^{-1}]^n$, which consists of $n \times 1$ column vectors with Laurent polynomials in t as entries. The vectors

$$v_{nk+j} \;=\; t^{-k}\, u_j \qquad (9.5)$$

form a basis of $\mathbb{C}[t, t^{-1}]^n$ (over $\mathbb{C}$) indexed by $\mathbb{Z}$. Thus we obtain an identification of $\mathbb{C}[t, t^{-1}]^n$ with $\mathbb{C}^\infty$. From (9.2) and (9.5),

$$e_{ij}(k)\, v_{ns+j} \;=\; v_{n(s-k)+i} \;. \qquad (9.6)$$

For $a(t) \in \widetilde{g\ell}_n$ we shall denote the corresponding matrix in $\overline{a}_\infty$ by $\tau(a(t))$. Then from (9.6) we can deduce that $e_{ij}(k)$ has, in the notation of Lecture 4 (see (4.3)), the following matrix representation in $\overline{a}_\infty$:

$$\tau(e_{ij}(k)) \;=\; \sum_{s \in \mathbb{Z}} E_{n(s-k)+i,\, ns+j} \;\cdot \qquad (9.7)$$

More generally, given $a(t) \in \widetilde{g\ell}_n$ as in (9.1), the corresponding matrix in $\overline{a}_\infty$ has the following block form:

$$\tau(a(t)) \;=\; \begin{bmatrix} \cdots\cdots\cdots\cdots\cdots\cdots\cdots\cdots\cdots \\ \cdots a_{-1} & a_0 & a_1 & \cdots\cdots\cdots \\ \cdots\cdots\cdots a_{-1} & a_0 & a_1 \cdots\cdots \\ \cdots\cdots\cdots\cdots a_{-1} & a_0 \cdots\cdots \\ \cdots\cdots\cdots\cdots\cdots\cdots\cdots\cdots\cdots \end{bmatrix} \qquad (9.8)$$

We regard (9.8) as a matrix of $\overline{a}_\infty$ in which the elements on each diagonal parallel to the principal diagonal form a periodic sequence with period n.

We note some properties of the mapping τ in the following proposition:

Proposition 9.1. (a) τ is an injective homomorphism of associative algebras, and hence Lie algebras, since the Lie bracket is the commutator.

(b) The image of $a(t) = \sum_j a_j t^j$ under τ is a strictly upper triangular matrix if and only if

$$a(t) = a_0 + a_1 t + a_2 t^2 + \ldots \text{ with } a_0 \text{ strictly upper triangular.}$$
(9.9)

(c) The shift operator Λ_j is the image under τ of $(a + tb)^j$, where

$$a = \sum_{i=1}^{n-1} e_{i,i+1}, \quad b = e_{n1}.$$

(d) Let $X(k) = t^k X$ be an element of $\widetilde{g\ell}_n$, where X is in $g\ell_n$. Define an anti-linear anti-involution ω on $\widetilde{g\ell}_n$ by

$$\omega(X(k)) = t^{-k} X^\dagger,$$
(9.10)

where $X^\dagger$ denotes the Hermitian adjoint of the $n \times n$ matrix X. Then

$$\tau(\omega(X(k))) = (\tau(X(k)))^\dagger,$$
(9.11)

where the symbol $\dagger$ on the right-hand side indicates the matrix Hermitian adjoint in $\bar{a}_\infty$.

Proof is straightforward. Let us check, for example, (c). Note that by (a):

$$\tau((a + bt)^j) = \left(\sum_{i=1}^{n-1} \tau(e_{i,i+1}) + \tau(te_{n1}) \right)^j$$

$$= \left(\sum_{i=1}^{n-1} \sum_s E_{ns+i,\, ns+i+1} + \sum_s E_{ns+n,\, n(s+1)+1} \right)^j$$

$$= \left(\sum_s \sum_{i=1}^{n} E_{ns+i,\, ns+i+1} \right)^j = \Lambda_1^j = \Lambda_j. \qquad \blacksquare$$

Remark 9.2. Viewing $\widetilde{g\ell}_n$ as a loop algebra, $\omega(a(t))$ is simply a pointwise Hermitian adjoint of the loop $a(t)$. Thus the corresponding "compact form"

$$\{ a(t) \in \widetilde{g\ell}_n \mid \omega(a(t)) = - a(t) \}$$

is simply the Lie algebra of maps (with finite Fourier series) of S^1 into su_n.

9.2. Representations of $\widehat{g\ell}'_n$ in $F^{(m)}$.

In the previous section we have given a realization of $\widetilde{g\ell}_n$ as a subalgebra of $\overline{a}_\infty$. As such it will have a projective representation in the wedge space $F^{(m)}$ (see §4.4) and its central extension

$$\widehat{g\ell}'_n = \widetilde{g\ell}_n \oplus \mathbb{C}c \tag{9.12}$$

will have a linear representation as a subalgebra of a_∞, the central extension of $\overline{a}_\infty$. We can compute the two-cocycle α on a pair of basis elements of $\widetilde{g\ell}_n$ of the form (9.2), using the representation (9.7) and (4.53), and we find

$$\alpha(\tau(e_{ij}(k)), \tau(e_{pq}(\ell))) = \delta_{iq} \delta_{jp} \delta_{\ell+k,0} k . \tag{9.13}$$

It now follows by linearity that if $X(k) = t^k X$, $Y(\ell) = t^\ell Y$ then

$$\alpha(\tau(X(k)), \tau(Y(\ell))) = \delta_{k,-\ell} k \operatorname{tr}(XY) , \tag{9.14a}$$

where tr denotes the trace in $g\ell_n$. For general elements $a(t)$, $b(t)$ in $\widetilde{g\ell}_n$ the formula (9.14a) can be written as follows:

$$\alpha(\tau(a(t)), \tau(b(t))) = \operatorname{Res}_0 \operatorname{tr} a'(t)b(t) \tag{9.14b}$$

where $a'(t)$ is the derivative of a with respect to t and Res_0 is the residue at $t = 0$, i.e., the coefficient of $1/t$.

We have thus been led by our search for highest weight representations of $\widetilde{g\ell}_n$ to consider its central extension $\widehat{g\ell}'_n$.

Definition 9.2. The Lie algebra $\widehat{g\ell}'_n$, defined by (9.12) and the commutation relations

$$[a(t), c] = 0 ,$$

$$[a(t), b(t)] = a(t)b(t) - b(t)a(t) + (\operatorname{Res}_0 \operatorname{tr} a'(t)b(t))c , \tag{9.15}$$

is called an *affine Kac-Moody algebra*, or simply an *affine algebra, associated to* $g\ell_n$.

We shall frequently use the commutation relations (9.15) for the elements $X(k) = t^k X$, $Y(m) = t^m Y$:

$$[X(k), Y(m)] = [X, Y](k+m) + k\delta_{k,-m}(\text{tr } XY)c. \qquad (9.16)$$

9.3. The invariant bilinear form on $\hat{g\ell}_n$. The action of $\widetilde{GL}_n$ on $\hat{g\ell}_n$.

Definition 9.3. A bilinear form $(\cdot \mid \cdot)$ on a Lie algebra g with Lie bracket $[\cdot, \cdot]$ is *invariant* if

$$([x, y]\mid z) = (x \mid [y, z]) \qquad \text{for all } x, y, z \in g \ . \qquad (9.17)$$

The bilinear form on $g\ell_n$ defined by

$$(X \mid Y) = \text{tr}(XY) \qquad (9.18)$$

is symmetric, non-degenerate and invariant, because of the properties of the trace. Now $g\ell_n$ is the Lie algebra of the group GL_n and $(\cdot \mid \cdot)$ has the property of being invariant under the adjoint action Ad of this group:

$$(\text{Ad}(A)(X) \mid \text{Ad}(A)(Y)) \equiv (AXA^{-1} \mid AYA^{-1}) = (X \mid Y) \qquad \text{for all } A \in GL_n \ . \qquad (9.19)$$

(Of course (9.17) is the infinitesimal equivalent for $g\ell_n$ of (9.19).)

We can define a bilinear form on $\widetilde{g\ell}_n$ in analogy with (9.18):

$$(X(k) \mid Y(m)) = \delta_{k+m,0} \text{ tr}(XY) \ . \qquad (9.20a)$$

This definition extends by linearity to general elements $a(t), b(t)$ of $\widetilde{g\ell}_n$ as follows:

$$(a(t) \mid b(t)) = \text{Res}_0 \ t^{-1} \text{ tr}(a(t)b(t)) \qquad (9.20b)$$

It is easily checked that $(\cdot \mid \cdot)$ is a symmetric, invariant, nondegenerate, bilinear form on $\widetilde{g\ell}_n$. It also has the property, which is analogous to (9.19), of being invariant under the adjoint action Ad of the group $\widetilde{GL}_n$, where

$$\widetilde{GL}_n \equiv GL_n(\mathbb{C}[t, t^{-1}]) \qquad (9.21)$$

is the group of all invertible $n \times n$ matrices over $\mathbb{C}[t, t^{-1}]$, *viz.* for all $A(t) \in \widetilde{GL}_n$ we have

$$(A(t)a(t)A^{-1}(t) \mid A(t)b(t)A^{-1}(t)) = (a(t) \mid b(t)) . \qquad (9.22)$$

The form $(\cdot \mid \cdot)$ can be extended to $\widehat{g\ell}_n'$ simply by defining

$$(c \mid \widetilde{g\ell}_n) = 0, \quad (c \mid c) = 0 . \qquad (9.23)$$

This definition preserves all the previous properties, except that now it is, of course, degenerate.

Definition 9.4. It is convenient from several points of view to enlarge $\widehat{g\ell}_n'$ by adding one more generator d:

$$\widehat{g\ell}_n = \widehat{g\ell}_n' \oplus \mathbb{C}d , \qquad (9.24)$$

where the commutation relations with the new generator d are:

$$[d, c] = 0, \quad [d, X(k)] = kX(k), \quad \text{i.e. } [d, a(t)] = t \, a'(t) . \qquad (9.25)$$

As before, $a'(t) = da/dt$. The Lie algebra $\widehat{g\ell}_n$ is also called an affine (Kac-Moody) algebra.

Proposition 9.2. The affine algebra $\widehat{g\ell}_n = \widetilde{g\ell}_n \oplus \mathbb{C}c \oplus \mathbb{C}d$ carries a nondegenerate, symmetric, invariant bilinear form $(\cdot \mid \cdot)$ defined by

$$\left. \begin{aligned} (a(t) \mid b(t)) &= \text{Res}_0 \; t^{-1} \text{tr}(a(t)b(t)) \quad \text{for } a(t), b(t) \in \widetilde{g\ell}_n \\[2mm] (c \mid a(t)) &= 0, \quad (c \mid c) = 0 , \\[2mm] (d \mid a(t)) &= 0, \quad (d \mid c) = 1, \quad (d \mid d) = 0 . \end{aligned} \right\} \qquad (9.26)$$

Proof. It is clear that the condition $(d \mid c) = 1$ ensures that $(\cdot \mid \cdot)$ is nondegenerate. Since $[c, d] = 0$, to prove invariance it suffices to show that for all $X(k), Y(m) \in \widehat{g\ell}_n$,

$$([X(k), d] \mid Y(m)) = (X(k) \mid [d, Y(m)]) .$$

This, however, follows immediately from (9.25) and (9.20a). ∎

Since, by (9.25), $dA - Ad = tA'$, we obtain

$$AdA^{-1} = d - tA'A^{-1} \tag{9.27}$$

for $A \in \widetilde{GL}_n$ acting on $\widetilde{g\ell}_n \oplus \mathbb{C}d$.

Now we want to lift the action of $\widetilde{GL}_n$ from $\widetilde{g\ell}_n \oplus \mathbb{C}d$ to $\widehat{g\ell}_n$. It is clear from the commutation relations of $\widehat{g\ell}_n$, viz. (9.15) and (9.25), that we must have:

$$\mathrm{Ad}(A(t))\,(c) = c ,$$

$$\mathrm{Ad}(A(t))\,(x(t)) = Ax(t)A^{-1} + \lambda(A,x)c ,$$

$$\mathrm{Ad}(A(t))\,(d) = d - tA'A^{-1} + \mu(A)c ,$$

where $A(t) \in \widetilde{GL}_n, x(t) \in \widetilde{g\ell}_n$ and $\lambda(A,x), \mu(A) \in \mathbb{C}$, and we used (9.27) in the last line. We demand further the $\widetilde{GL}_n$ -invariance of the form $(\cdot \mid \cdot)$ on $\widehat{g\ell}_n$. This gives us

$$0 = (x \mid d) = (\mathrm{Ad}(A)(x) \mid \mathrm{Ad}(A)(d))$$

$$= (AxA^{-1} + \lambda c \mid d - tA'A^{-1} + \mu c) ,$$

from which $\lambda = \mathrm{Res}_0 \ \mathrm{tr}\, A'xA^{-1}$. Similarly, from

$$0 = (d \mid d) = (\mathrm{Ad}(A)(d) \mid \mathrm{Ad}(A)(d))$$

we get $\mu = -\tfrac{1}{2} \mathrm{Res}_0 \ \mathrm{tr}(t(A'A^{-1})^2)$.

Summarizing:

$$\mathrm{Ad}(A(t))\,(c) = c , \tag{9.28a}$$

$$\mathrm{Ad}(A(t))\,(x(t)) = Ax(t)A^{-1} + \mathrm{Res}_0\,\mathrm{tr}(A'x(t)A^{-1})c , \tag{9.28b}$$

$$\mathrm{Ad}(A(t))\,(d) = d - tA'A^{-1} - \frac{1}{2}\mathrm{Res}_0\,\mathrm{tr}(t(A'A^{-1})^2)c . \tag{9.28c}$$

One checks immediately that these formulas indeed define automorphism, of the Lie algebra $\hat{g\ell}_n$.

9.4. Reduction from a_∞ to $\hat{s\ell}_n$ and the unitarity of highest weight representations of $\hat{s\ell}_n$.

We showed in §9.1 (see Proposition 9.1) that $\widetilde{g\ell}_n$ is a subalgebra of $\bar{a}_\infty$ and that the antilinear anti-involution ω on $\widetilde{g\ell}_n$ coincides with the one induced from $\bar{a}_\infty$. The central extension $\widetilde{g\ell}'_n$ of $\widetilde{g\ell}_n$ is a subalgebra of a_∞ and if we put in addition $\omega(c) = c$, the antilinear anti-involutions on $\widetilde{g\ell}'_n$ and a_∞ are consistent. Moreover, $\widetilde{g\ell}'_n$ contains the principal subalgebra $\mathscr{A}$ of a_∞ (recall Proposition 9.1 (c)). We know from Lecture 4 that a_∞ has a sequence of fundamental irreducible representations $\hat{r}_m$ in $F^{(m)}$, which remain irreducible when restricted to the subalgebra $\mathscr{A}$ (see Section 5.1). Thus $\widetilde{g\ell}'_n$ has a sequence of irreducible representations $\hat{\pi}_m$ in $F^{(m)}$ and for $a(t)$ as in (9.9) we have:

$$\hat{\pi}_m \left(\tau(a(t)) \right) \psi_m = 0 . \tag{9.29}$$

Moreover, $\hat{\pi}_m(c)$ acts by 1 in each $F^{(m)}$, while the action of the diagonal elements

$$\hat{\pi}_m \left(\tau(e_{ii}(0)) \right) \psi_m = \sum_{s \in \mathbb{Z}} \hat{r}_m(E_{ns+i,\, ns+i}) \psi_m \tag{9.30}$$

can be determined from:

$$\hat{r}_m(E_{ns+i,\, ns+i}) \psi_m = \lambda(ns + i, m) \psi_m \tag{9.31}$$

where

$$\lambda(j,\, m) = \begin{cases} 1 & \text{if } m \geqslant j > 0 \\ -1 & \text{if } 0 \geqslant j > m \\ 0 & \text{otherwise} . \end{cases}$$

It is easy to see (e.g. by Corollary 10.1 (a) in Lecture 10) that the representations $\hat{\pi}_m$ of $\widetilde{g\ell}'_n$ extend uniquely to $\hat{g\ell}_n$ (so that (9.25) holds) subject to the condition

$$\pi_m(d)\psi_m = 0 . \tag{9.32}$$

The Lie algebra $\mathfrak{sl}_n$ of $n \times n$ complex traceless matrices is a Lie subalgebra of $\mathfrak{gl}_n$. Hence $\widetilde{\mathfrak{sl}}_n$, $\mathfrak{sl}_n'$ and $\hat{\mathfrak{sl}}_n$ can be defined and are Lie subalgebras of $\widetilde{\mathfrak{gl}}_n$, $\mathfrak{gl}_n'$ and $\hat{\mathfrak{gl}}_n$ respectively. Consequently $\hat{\mathfrak{sl}}_n'$ is also a subalgebra of a_∞. The Cartan subalgebra h of $\mathfrak{sl}_n$ has as basis the diagonal traceless matrices $e_{ii} - e_{i+1,i+1}$ $(1 \leqslant i \leqslant n - 1)$. The *Cartan subalgebra* $\hat{h}$ of $\hat{\mathfrak{sl}}_n$ is spanned by

$$\{ h_i \equiv e_{ii}(0) - e_{i+1,i+1}(0) \quad (1 \leqslant i \leqslant n - 1); \; c; \; d \} \; .$$

We shall, however, choose as basis

$$\{ h_0 \equiv c + e_{n,n}(0) - e_{1,1}(0); \; h_i \; (1 \leqslant i \leqslant n - 1); \; d \} \; .$$

Define linear functionals ω_j $(j = 0, 1, \ldots, n - 1)$ on $\hat{h}$ by

$$\omega_j(h_i) = \delta_{ij} \; (0 \leqslant i, j \leqslant n - 1); \quad \omega_j(d) = 0.$$

Using (9.30) and (9.31) we can verify that

$$\hat{\pi}_m(h_i) \psi_m = \omega_{m'}(h_i) \psi_m \; , \tag{9.33a}$$

or, more generally, using (9.32):

$$\hat{\pi}_m(h) \psi_m = \omega_{m'}(h) \psi_m \quad \text{for } h \in \hat{h} \; , \tag{9.33b}$$

where m' denotes the number from $\{ 0, 1, \ldots, n - 1 \}$ congruent to $m \bmod n$.

Furthermore, denoting by n_+ (resp. n_-) the subalgebras of the strictly upper (resp. strictly lower) triangular matrices of $\mathfrak{sl}_n$, we have the triangular decomposition: $\mathfrak{sl}_n = n_- \oplus h \oplus n_+$. The corresponding *triangular decomposition* of $\hat{\mathfrak{sl}}_n$ is constructed as follows. Put $\hat{n}_+ = n_+ + \sum_{k > 0} t^k \mathfrak{sl}_n$, $\hat{n}_- = n_- + \sum_{k > 0} t^{-k} \mathfrak{sl}_n$. Then we have:

$$\hat{\mathfrak{sl}}_n = \hat{n}_+ \oplus \hat{h} \oplus \hat{n}_- \; .$$

Definition 9.5. For a given $\lambda \in \hat{h}^*$, called the *highest weight*, we define the *highest weight representation* π_λ of the Lie algebra $\hat{\mathfrak{sl}}_n$ as an irreducible representation on a vector space $L(\lambda)$ which admits a non-zero vector v_λ, called a *highest weight vector*, such that

$$\pi_\lambda(\hat{n}_+)v_\lambda = 0 \ ,$$

$$(9.34)$$

$$\pi_\lambda(h)v_\lambda = \lambda(h)v_\lambda \quad \text{for } h \in \hat{h} \ .$$

Remark 9.3. A general argument (as used in Lecture 3 for the Virasoro algebra) proves the existence and uniqueness of $L(\lambda)$ for all $\lambda \in \hat{h}^*$.

We see that we have a representation of $\hat{sl}_n$ in $F^{(m)}$ which, with $v_\lambda = \psi_m$, satisfies the requirements (9.34) for a highest weight representation (here we use Proposition 9.1(b)). This representation is unitary due to Proposition 9.1(d) and the unitarity of the representation of a_∞ in each $F^{(m)}$.

Recall that the representation $F^{(m)}$ is irreducible under $\hat{gl}'_n$ since the latter contains the $\Lambda_j (j \in \mathbb{Z})$, i.e., the generators of the oscillator algebra $\mathscr{A}$. To determine whether Λ_j belongs to the traceless subalgebra $\hat{sl}'_n$ of $\hat{gl}'_n$, we recall from Proposition 9.1(c) that Λ_j is the image in a_∞ of the j-th power of the $n \times n$ matrix

$$a = \begin{bmatrix} 0 & 1 & 0 & \ldots\ldots & 0 \\ 0 & 0 & 1 & \ldots\ldots & 0 \\ \cdot & \cdot & \cdot & \cdot\cdot\cdot\cdot\cdot\cdot\cdot & \cdot \\ \cdot & \cdot & \cdot & \cdot\cdot\cdot\cdot\cdot\cdot\cdot & \cdot \\ 0 & 0 & 0 & \ldots\ldots & 1 \\ t & 0 & 0 & \ldots\ldots & 0 \end{bmatrix}$$

It is easily checked that a^j is traceless when j is not an integral multiple of n, while $a^{sn} = t^s I$. Hence $\Lambda_j \in \hat{sl}'_n$ when j is not an integral multiple of n, while $\hat{sl}'_n$ commutes with all Λ_j for which j is an integral multiple of n. Since the subspace

$$F^{(m)}_{(0)} = \{ v \in F^{(m)} \mid \hat{r}(\Lambda_{sn})v = 0, \ s = 1, 2, \ldots \} \tag{9.35}$$

is spanned by the vectors obtained by the repeated applications on ψ_m of the Λ_{-j} for $j \neq sn$ ($s \in \mathbb{N}$), it is an irreducible invariant subspace for $\hat{sl}'_n$. It follows immediately that $F^{(m)}_{(0)}$ is invariant and irreducible also under $\hat{sl}_n$.

We can determine $F^{(m)}_{(0)}$ in more explicit form with the help of the boson-fermion equivalence under which $F^{(m)}$ is isomorphic to $B^{(m)} = \mathbb{C}[x_1, x_2, \ldots]$.

The image $B_{(0)}^{(m)}$ of $F_{(0)}^{(m)}$ is easily determined since Λ_j is represented by $\partial/\partial x_j$ for $j > 0$. Clearly

$$B_{(0)}^{(m)} = \mathbb{C}[\, x_j \,|\, j \geqslant 1, \quad j \text{ is not a multiple of } n \,] \ ,$$

where on the right-hand side we have the subspace of polynomials in those variables x_j for which j is not a multiple of n.

Thus, $F_{(0)}^{(m)} = B_{(0)}^{(m)}$ is a unitary highest weight representation space of $\hat{s\ell}_n$ with highest weight $\omega_{m'}$, where $m' \in \{0,1,\ldots,n-1\}$ and m' is congruent to m modulo n, i.e. $F_{(0)}^{(m)} \cong L(\omega_{m'})$. The representation $L(\omega_m)\,(0 \leqslant m \leqslant n-1)$ is called the m-th *fundamental representation* of $\hat{s\ell}_n$.

We can clearly take tensor products of the fundamental representations and the highest component will have a highest weight equal to the sum of the individual highest weights. Summarizing, we have proved:

Proposition 9.4. The representations

$$L(k_0 \omega_0 + k_1 \omega_1 + \ldots + k_{n-1} \omega_{n-1})$$

of $\hat{s\ell}_n$, where $k_i \in \mathbb{Z}_+$, $0 \leqslant i \leqslant n-1$, are unitary. ∎

Remark 9.4. Proposition 9.4 is a special case of a theorem of Garland [1978] and Kac-Peterson [1984b] for non-twisted affine algebras and general Kac-Moody algebras respectively.

Theorem 9.1. The representation $L(\lambda)$ of $\hat{s\ell}_n$ is unitary if and only if $\lambda(h_i) \in \mathbb{Z}_+$ for $i = 0, \ldots, n-1$ and $\lambda(d) \in \mathbb{R}$.

Proof. The 'if' part follows from Proposition 9.4 since we can make $\lambda(d) = 0$ by adding to d an arbitrary real constant. To see the 'only if' part, put

$$e_0 = e_{n,1}(1), \quad f_0 = e_{1,n}(-1) \ ,$$

$$e_i = e_{i,i+1}(0), \quad f_i = e_{i+1,i}(0) \quad \text{for } i = 1,\ldots,n-1 \ .$$

Then $\{\, e_i, h_i, f_i \,\}$ form a standard basis of $s\ell_2$ for each i, and $\omega(e_i) = f_i$ (see (9.10)). But the only unitary irreducible representations of $s\ell_2$ with this involution are the finite-dimensional ones. Thus, unitarity of $L(\lambda)$ implies that $\lambda(h_i) \in \mathbb{Z}_+$ for all i, and, in particular, that $\lambda(c) \in \mathbb{Z}_+$. ∎

Since $c = h_0 + \ldots + h_{n-1}$, we deduce

Corollary 9.1. If the representation $L(\lambda)$ of $\widehat{s\ell}_n$ is unitary, then $\lambda(c) \in \mathbb{Z}_+$. ∎

The integer $\lambda(c)$ is called the *level* of $L(\lambda)$.

LECTURE 10

10.1. Nonabelian generalization of Virasoro operators: the Sugawara construction.

In Lecture 9 we constructed the affine algebra $\hat{g\ell}'_n$ starting from the finite dimensional Lie algebra $g\ell_n$. The restriction of the bilinear form (9.18) on $g\ell_n$ to its subalgebra $s\ell_n$ remains nondegenerate and we have the associated affine algebra $\hat{s\ell}'_n$. In fact for any finite-dimensional Lie algebra g which has an invariant symmetric nondegenerate bilinear form $(\cdot \mid \cdot)$ there is a corresponding affine algebra $\hat{g}' = \tilde{g} \oplus \mathbb{C}c$, with commutation relations

$$[x(k), c] = 0,$$

$$[x(k), y(m)] = [x, y](k + m) + k\delta_{k, -m}(x \mid y)c,$$

(10.1)

in analogy with (9.16). In the special case where g is the one-dimensional abelian Lie algebra, (10.1) evidently reduces to the commutation relations (2.2) of the oscillator algebra $\mathscr{A}$. Thus we can view $\hat{g}'$ as a nonabelian generalization of $\mathscr{A}$.

We define $\hat{g} = \hat{g}' \oplus \mathbb{C}d$ in the same way as in Definition 9.4 for $g = s\ell_n$, and, as in §9.4, introduce the Cartan subalgebra $\hat{h} = h \oplus \mathbb{C}c \oplus \mathbb{C}d$, where h is a Cartan subalgebra of g. Then $\hat{g}$ carries a nondegenerate symmetric invariant bilinear form $(\cdot \mid \cdot)$ defined by (9.26), with tr ab replaced by $(a \mid b)$.

An important example of g which has an invariant symmetric nondegenerate bilinear form is a *reductive* Lie algebra (= a direct sum of simple Lie algebras and an abelian Lie algebra). Then one has a triangular decomposition of g, and in the same way as for $\hat{s\ell}_n$ in Section 9.4 one defines the associated triangular decomposition of $\hat{g}$ and irreducible highest weight representations $L(\lambda)$. It is easy to show that $L(\lambda)$ remains irreducible when restricted to $\hat{g}'$ (see e.g. Kac [1983], §9.10). In what follows, g is assumed to be a (finite-dimensional) reductive Lie algebra.

105

We shall show in this lecture that we can use the generators of $\hat{g}'$ to construct representations of *Vir* in a nonabelian generalization of Virasoro's construction described in Lecture 2. Before doing this we shall need to collect some simple properties of g and $\hat{g}'$.

Let $\{\,u_i \mid i = 1, \ldots, \dim g\,\}$ be a basis in g and let $\{u^i \mid i = 1, \ldots, \dim g\}$ be the dual basis, so that

$$(u_i \mid u^j) = \delta_{ij} \ .$$

Then for any $x \in g$ we have

$$x = (x \mid u_i)u^i = (x \mid u^i)u_i \ . \tag{10.2}$$

Note that in (10.2) and subsequently we have adopted the convention that a repeated upper and lower index is summed over the dimension of g.

Let $\Omega_0 = u_i u^i$ be the *Casimir operator* of g. It is easily seen to be independent of the choice of dual bases $\{u_i\}$ and $\{u^i\}$. In particular, we have

$$[u_i, u^i] = 0 \ . \tag{10.3}$$

Lemma 10.1. $\ [\,g, \Omega_0\,] = 0.$

Proof. Let $x \in g$. Then,

$$[x, u_i u^i] = [x, u_i]u^i + u_j[x, u^j]$$

$$= [x, u_i]u^i + u_j([x, u^j] \mid u_i)u^i \qquad \text{(by (10.2))}$$

$$= [x, u_i]u^i - u_j(u^j \mid [x, u_i])u^i \qquad \text{(by invariance)}$$

$$= [x, u_i]u^i - [x, u_i]u^i \qquad \text{(by (10.2))}$$

$$= 0 \ . \qquad\qquad\qquad\qquad \blacksquare$$

The Casimir operator Ω_0 acts as a multiplication by a scalar on any highest weight representation V of g with highest weight λ (due to Lemma 10.1). Recall that one has:

$$\Omega_0 = (\lambda \mid \lambda + 2\bar{\rho})I \ , \tag{10.4}$$

where $\bar{\rho}$ is the sum of the fundamental weights of g. (Recall that (10.4) is proved by choosing a basis $\{u_i\}$ of g consistent with the root space decomposition, replacing in Ω_0 expressions $e_\alpha e_{-\alpha}$ by $e_{-\alpha}e_\alpha + h_\alpha$ for $\alpha > 0$ and using the fact that $2\bar{\rho}$ is a sum of positive roots; then it is immediate that $\Omega_0(v_\lambda) = (\lambda \mid \lambda + 2\bar{\rho})v_\lambda$ if v_λ is a highest weight vector.)

Since

$$(u_i(m) \mid u^j(n)) = \delta_{ij}\,\delta_{m,\,-n}\,, \tag{10.5}$$

from (10.1) and (10.3) we have:

$$[u_i(m),\, u^i(n)] = m\,c\,\delta_{m,\,-n}\,\dim\,g\,. \tag{10.6}$$

Lemma 10.2. $[x, u_i]\,(m)\,u^i(n) + u_i(m)\,[x, u^i]\,(n) = 0$, for $x \in g$ and $m, n \in \mathbb{Z}$.

Proof is similar to that of Lemma 10.1. ∎

Lemma 10.3. Let the Lie algebra g be simple or abelian. Then,

$$[u_i,\, [u^i,\, x]] = 2gx = [u^i,\, [u_i,\, x]]\,,$$

where g is a scalar (the factor 2 is inserted for convenience) and $g = 0$ if g is abelian.

Proof. The lemma clearly holds if g is abelian. Since Ω_0 commutes with g, it acts as a scalar in every irreducible representation of g, in particular in the adjoint representation if g is simple. Denoting this scalar by $2g$ and noting that in the adjoint representation

$$\Omega_0 = (\text{ad } u_i)(\text{ad } u^i) = (\text{ad } u^i)(\text{ad } u_i)$$

completes the proof. ∎

Definition 10.1. A representation of the affine algebra $\hat{g}'$ on a vector space V is called *admissible* if for every $v \in V$ we have $x(k)(v) = 0$ for all $x \in g$ and all $k \gg 0$. It is clear that $L(\lambda)$ is an admissible representation of $\hat{g}'$.

Proposition 10.1. Let g be a simple or abelian Lie algebra with a nondegenerate symmetric bilinear invariant form $(\cdot \mid \cdot)$, and let $\hat{g}'$ be the corresponding affine algebra with commutation relations (10.1). Consider an admissible representation of $\hat{g}'$ on a vector space V. Then the operators T_k defined by

$$T_k = \frac{1}{2} \sum_{j \in \mathbb{Z}} : u_i(-j)\, u^i(j + k): \qquad\qquad (10.7)$$

for $k \in \mathbb{Z}$, satisfy the commutation relations

$$[\, x(n),\ T_k \,] = (c + g) n\, x(n + k)\,. \qquad\qquad (10.8)$$

Remark 10.1. The *normal ordering* : : in (10.7) means as usual that the order within the colons is to be preserved when $-j \leqslant j + k$ and reversed otherwise. Then the series in (10.7) will always be finite when applied to any $v \in V$ (for admissible V). Also, since $u_i(m)u^i(n)$ is independent of the choice of dual bases, this holds for the T_k as well. Finally, it follows from (10.6) that we may drop the sign of normal ordering in (10.7) if $k \neq 0$.

Proof of Proposition 10.1. We shall use the cutoff procedure described in Lecture 2. Let

$$T_k(\epsilon) = \frac{1}{2} \sum_{j \in \mathbb{Z}} : u_i(-j)\, u^i(j + k): \ \psi(\epsilon j)\,,$$

where $\psi(x)$ is the cutoff function defined in (2.13a). Then

$$[\, x(n),\ T_k(\epsilon)\,] = \frac{1}{2} \sum_j [\, x,\ u_i\,](n - j)\, u^i(j + k)\psi(\epsilon j)$$

$$+ \frac{1}{2} \sum_j u_i(-j)[\, x,\ u^i\,](j + k + n)\psi(\epsilon j) + \frac{1}{2}\, nc\, x(n + k)\psi(\epsilon n)$$

$$+ \frac{1}{2}\, nc\, x(n + k)\psi(\epsilon(n + k))\,. \qquad\qquad (10.9)$$

We split the first sum in (10.9) into terms for which $j \geqslant (n - k)/2$ (hence in normal order) and those for which $j < (n - k)/2$. The latter are replaced by normal-ordered terms with the help of the commutation relation

$$[[\, x,\ u_i\,](m),\ u^i(n)\,] = [[\, x,\ u_i\,],\ u^i\,](m + n)\,, \qquad (10.10a)$$

which follows immediately from (10.1), since the coefficient of c vanishes by the invariance of the bilinear form and (10.3). We now apply Lemma 10.3

to the right-hand side of (10.10a). The second sum in (10.9) is similarly split into normal-ordered terms satisfying $j \geqslant -(n+k)/2$ and the remainder for which $j < -(n+k)/2$. For the latter we use the dual of (10.10a):

$$[[x, u^i](m), u_i(n)] = [[x, u^i], u_i](m+n) \qquad (10.10b)$$

and Lemma 10.3 once more. In this way we get

$$[x(n), T_k(\epsilon)] = \frac{1}{2}\sum_j \: : [x, u_i](n-j)u^i(j+k): \: \psi(\epsilon j)$$

$$+ \frac{1}{2}\sum_j : u_i(-j)[x, u^i](j+k+n): \psi(\epsilon j) + \frac{1}{2} nc\, x(n+k)\psi(\epsilon n)$$

$$+ \frac{1}{2} nc\, x(n+k)\psi(\epsilon(n+k)) + gx(n+k)\sum_j{}' \psi(\epsilon j) \; ,$$

where Σ' is taken over $-(n+k)/2 \leqslant j < (n-k)/2$. Making the transformation $j \to j + n$ in the first sum, we obtain in the limit $\epsilon \to 0$:

$$[x(n), T_k] = \frac{1}{2}\sum_j : [x, u_i](-j)u^i(j+k+n)$$

$$+ u_i(-j)[x, u^i](j+k+n): \: + n(g+c)x(n+k).$$

By Lemma 10.2 each term in the sum vanishes, which proves the proposition. ∎

Theorem 10.1. Under the same hypotheses as in Proposition 10.1 we have

$$[T_n, T_k] = (c+g)(n-k)T_{n+k} + \delta_{n,-k}\frac{(n^3-n)}{12} (\dim \mathfrak{g})c(c+g) \; . \qquad (10.11)$$

Proof.

$$[T_n(\epsilon), T_k] = \frac{1}{2}\sum_{j \in \mathbb{Z}} [u_i(-j)u^i(j+n), T_k]\psi(\epsilon j)$$

$$= \frac{1}{2}(c+g) \sum_j (-j)u_i(k-j)u^i(j+n)\psi(\epsilon j)$$

$$+ \frac{1}{2}(c+g) \sum_j (j+n)u_i(-j)u^i(j+k+n)\psi(\epsilon j)$$

by Proposition 10.1. We now reorder terms with the help of (10.6) to get normal-ordered expressions:

$$[T_n(\epsilon), T_k] = \frac{1}{2}(c+g) \sum_j (-j) : u_i(k-j)u^i(j+n) : \psi(\epsilon j)$$

$$+ \frac{1}{2}(c+g) \sum_j (j+n) : u_i(-j)u^i(j+k+n) : \psi(\epsilon j)$$

$$+ \frac{1}{2}c(c+g)(\dim g)\,\delta_{n,-k}\left\{ -\sum_{j=-1}^{-n} j(n+j)\psi(\epsilon j) \right\} .$$

Making the transformation $j \to j+k$ in the first sum, we obtain (10.11) on taking the limit $\epsilon \to 0$. ∎

Remark 10.2. Let g be simple. We normalize the bilinear form $(\cdot\,|\,\cdot)$ by choosing long roots to have square length 2. (Note that for $\bar{g} = \mathfrak{sl}_n$, $(x\,|\,y) = \operatorname{tr} xy$.) With this normalization, it can be shown in the same way as for $\mathfrak{sl}_n$, that if $L(\Lambda)$ is a unitary highest weight representation of $\hat{g}$ then the central element c is represented by mI, where m is a non-negative integer. In this normalization of $(\cdot\,|\,\cdot)$, the number g, which is half the eigenvalue of Ω_0 in the adjoint representation, is a positive integer. It is known as the *dual Coxeter number*. Using (10.4) for $\lambda = \theta$, the highest root, we find:

$$g = 1 + (\theta\,|\,\bar{\rho}) .$$

The concrete values of g are given below.

Table 10.1

Simple Lie algebra g	dim g	Dual Coxeter number g
A_ℓ	$\ell^2 + 2\ell$	$\ell + 1$
B_ℓ	$2\ell^2 + \ell$	$2\ell - 1$
C_ℓ	$2\ell^2 + \ell$	$\ell + 1$
D_ℓ	$2\ell^2 - \ell$	$2\ell - 2$
E_6	78	12
E_7	133	18
E_8	248	30
F_4	52	9
G_2	14	4

Proposition 10.1 and Theorem 10.1 now give us:

Corollary 10.1. Let V be an admissible representation of $\hat{g}'$ (g is simple or abelian) such that $c = mI$ with $m \neq -g$. Then:
(a) We can define

$$L_k = \frac{1}{(m+g)} \ T_k \tag{10.12}$$

so that we have:

$$[L_k, L_n] = (k-n)L_{k+n} + \delta_{k+n,0} \frac{k^3 - k}{12} \frac{m \dim g}{m+g} \ , \tag{10.13a}$$

$$[L_k, x(n)] = -nx(n+k) \ . \tag{10.13b}$$

In particular, putting $d = -L_0$, V can be extended to a representation of $\hat{g}$.
(b) If V is unitary, we thus obtain a unitary representation of *Vir* in V with central charge

$$c = \frac{m \dim g}{m+g} \ . \tag{10.14}$$

Proof. The only thing that remains to check is unitarity. For that choose a basis $\{v_j\}$ of the compact form of g (= the fixed point set of $-\omega$, where ω is the compact antilinear anti-involution of g) such that $(v_j \mid v_k) = -\delta_{jk}$, and put $u_j = u^j = i v_j$. Then.

$$\omega(L_n) = \omega\left(\sum_i \sum_{j \in \mathbb{Z}} u_i(-j)u_i(j+n)\right) = \sum_i \sum_{j \in \mathbb{Z}} u_i(-j-n)u_i(j) = L_{-n}$$

if $n \neq 0$ (see Remark 10.1). A similar calculation applies in the case $n = 0$. ∎

Remark 10.3. Note from Table 10.1 that $c \geqslant 1$ in (10.14).

Let

$$g = \overset{\ell}{\underset{i=0}{\oplus}} g_i \qquad (10.15)$$

be a reductive Lie algebra with center g_0. We can choose bases $\{u_j\}$, $\{u^j\}$ in g which respect the decomposition (10.15) and are dual with respect to an invariant, symmetric bilinear form $(\cdot \mid \cdot)$. We assume that this form is properly normalized when restricted to each component of g (Remark 10.2). Note that $\hat{g}'$ is the direct sum of the $\hat{g}'_i$. Corresponding to the unitary highest weight representation of $\hat{g}'_i$ in $L(\Lambda_i)$ we have a family of operators $L_k^{(i)}(k \in \mathbb{Z})$ which satisfy the Virasoro algebra, where m_i is the level of $L(\Lambda_i)$ and g_i the dual Coxeter number of g_i. Furthermore, $\hat{g}'$ acts on the tensor product of the $L(\Lambda_i)$ in the usual way, so that the $L_k^{(i)}$ commute for different i. Hence, defining for each $k \in \mathbb{Z}$

$$L_k = \sum_{i=0}^{\ell} L_k^{(i)}, \qquad (10.16)$$

we get:

Corollary 10.2. The $L_k(k \in \mathbb{Z})$ form a unitary representation of the Virasoro algebra with central charge given by

$$c = \sum_{i=0}^{\ell} \frac{(\dim g_i)m_i}{m_i + g_i} \qquad (10.17)$$

Remark 10.4. The construction of Theorem 10.1 is a discrete counterpart of the Sugawara [1968] construction. The earliest reference that we know in which the central charge is given correctly (in the case of $su(n)$) is the paper Dashen-Frishman [1975]. In the following we shall follow standard practice and refer to this as the Sugawara construction.

Note that

$$T_0 = \frac{1}{2} u_i u^i + \sum_{j>0} u_i(-j)u^i(j) \ . \tag{10.18}$$

The operator T_0 is closely related to the Casimir operator Ω of $\hat{g}$ introduced by Kac [1974] in the framework of general Kac-Moody algebras:

Proposition 10.2. Let g be a simple finite-dimensional Lie algebra.

(a) The operator Ω defined by

$$\Omega = 2(c + g)d + 2T_0 \tag{10.19}$$

commutes with every element of $\hat{g}$.

(b) On $L(\Lambda)$ the eigenvalue of Ω is $(\Lambda + 2\rho \mid \Lambda)$ where $\rho = \sum_i \omega_i$ is the sum of the fundamental weights (which are defined in the same way as for $\hat{sl}_n$ in §9.4).

Proof. (a) We see that $[d, \Omega] = 0$ by using the representation (10.18) for T_0 and (9.25). The equality $[x(n), \Omega] = 0$ follows from the application of (9.25) and (10.8).

(b) Applying Ω to the highest weight vector we see from (9.29), (9.32) and (10.18) that the only nonzero contribution is from $2(c + g)d + \Omega_0$. Using (10.4), we get the formula immediately since $\rho = \bar{\rho} + g\omega_0$. ∎

10.2. The Goddard-Kent-Olive construction.

In the previous section (see Remark 10.3 and Corollary 10.2) we saw that the unitary representation of *Vir* obtained from $\hat{g}'$, where g is a reductive Lie algebra, has a central charge which is always greater than or equal to 1. Goddard-Kent-Olive [1985] have found a way to construct unitary representations of *Vir* for which the central charge is less than 1.

We take a reductive Lie algebra g and a reductive subalgebra p of g. Given a unitary highest weight representation of g we can use the Sugawara construction to form a unitary representation of *Vir*, viz. $\{L_k^g \mid k \in \mathbb{Z}\}$ with central charge c_g from the generators of $\hat{g}'$. We can construct a second unitary representations of *Vir* $\{L_k^p \mid k \in \mathbb{Z}\}$ with central charge c_p from the generators of $\hat{p}'$. (We have to be careful to follow the normalization of Remark 10.2 separately for $\hat{g}'$ and $\hat{p}'$).

Theorem 10.2. The operators

$$L_k = L_k^g - L_k^p \quad (k \in \mathbb{Z}) \tag{10.20}$$

form a unitary representation of the Virasoro algebra with central charge

$$c = c_g - c_p \ . \tag{10.21}$$

Proof. We first show that L_k commutes with L_n^p. This follows from the fact that L_k commutes with $x(n)$ when $x(n) \in \hat{p}'$:

$$[L_k, x(n)] = [L_k^g, x(n)] - [L_k^p, x(n)]$$

$$= nx(n+k) - nx(n+k) = 0 \ .$$

Thus

$$[L_k, \hat{p}'] = 0 \quad (k \in \mathbb{Z}) \ , \tag{10.22}$$

and hence

$$[L_k, L_n^p] = 0 \quad (k, n \in \mathbb{Z}) \ .$$

It follows now that

$$[L_k, L_n] = [L_k, L_n^g] = [L_k^g, L_n^g] - [L_k^p, L_n^g]$$

$$= [L_k^g, L_n^g] - [L_k^p, L_n + L_n^p]$$

$$= [L_k^g, L_n^g] - [L_k^p, L_n^p]$$

$$= (k-n)L_{k+n} + \delta_{k+n,0} \frac{(k^3 - k)}{12} (c_g - c_p) \ .$$

The unitarity of this representation of *Vir* follows from Corollary 10.1. ∎

We now consider the special case where p is a simple Lie algebra and $g = p \oplus p$. We consider two representations of $\hat{p}$ on $L(\Lambda)$ and $L(\Lambda')$ with levels m and m' respectively. Then the action of $\hat{g}' = \hat{p}' \oplus \hat{p}'$ on $L(\Lambda) \otimes L(\Lambda')$ is given by:

$$(x(n) \oplus y(m))(v \otimes w) = (x(n)(v)) \otimes w + v \otimes (y(m)(w))$$

for $v \otimes w \in L(\Lambda) \otimes L(\Lambda')$. From this it is clear that the Sugawara construction on $\hat{g}'$ gives

$$L_k^g = L_k^p \otimes 1 + 1 \otimes L_k^p \,,$$

with central charge $c_g = (\dim \, p \,)(m/(m+g) + m'/(m'+g))$. Also $\hat{p}'$ has the "diagonal" action on $L(\Lambda) \otimes L(\Lambda')$ given by

$$x(n) \cdot (v \otimes w) = (x(n)(v)) \otimes w + v \otimes (x(n)(w)) \,.$$

(This is equivalent to embedding $\hat{p}'$ diagonally in $\hat{p}' \oplus \hat{p}'$, i.e. $x(n) \to x(n) \oplus x(n)$.)) The level of this representation of $\hat{p}'$ is clearly $m + m'$. This gives rise to Virasoro operators L_k^p with central charge $= (\dim p)(m + m')/(m + m' + g)$. Hence, $L_k = L_k^g - L_k^p$ gives, by Theorem 10.2, a representation of *Vir* with central charge

$$c = (\dim \, p) \left(\frac{m}{m + g} + \frac{m'}{m' + g} - \frac{m + m'}{m + m' + g} \right) \,.$$

From (10.19) we observe that

$$L_0 = \left(\frac{1}{2(m+g)} \, \Omega^p \, \Big|_{L(\Lambda)} - d \right) \otimes 1 + 1 \otimes \left(\frac{1}{2(m'+g)} \, \Omega^p \, \Big|_{L(\Lambda')} - d \right)$$

$$- \left(\frac{1}{2(m+m'+g)} \, \Omega^p \, \Big|_{L(\Lambda) \otimes L(\Lambda')} - d \otimes 1 - 1 \otimes d \right) \,.$$

So by Proposition 10.2(b) we have

$$L_0 = \frac{1}{2} \left(\frac{(\Lambda \mid \Lambda + 2\rho)}{m + g} + \frac{(\Lambda' \mid \Lambda' + 2\rho)}{m' + g} - \frac{\Omega^p}{m + m' + g} \right) \,.$$

We summarize these results in the following proposition.

Proposition 10.3. Let p be a simple Lie algebra with dual Coxeter number g. Let $\{u_i\}$ and $\{u^i\}$ be dual bases of p, and consider two highest weight unitary representations $L(\Lambda), L(\Lambda')$ of $\hat{p}$ with levels m and m'. Then

(a) The following operators $(k \in \mathbb{Z})$

$$L_k = \left(\frac{1}{2(m+g)} - \frac{1}{2(m+m'+g)} \right) \sum_{j \in \mathbb{Z}} : u_i(-j)u^i(j+k) : \otimes 1$$

$$+ \left(\frac{1}{2(m'+g)} - \frac{1}{2(m+m'+g)} \right) \sum_{j \in \mathbb{Z}} 1 \otimes : u_i(-j)u^i(j+k) :$$

$$- \frac{1}{(m+m'+g)} \sum_{j \in \mathbb{Z}} u_i(-j) \otimes u^i(j+k) , \tag{10.23}$$

on the space $L(\Lambda) \otimes L(\Lambda')$ form a unitary representation of the Virasoro algebra with central charge

$$c = (\dim p)\left(\frac{m}{m+g} + \frac{m'}{m'+g} - \frac{m+m'}{m+m'+g} \right) . \tag{10.24}$$

(b) $$L_0 = \frac{1}{2}\left(\frac{(\Lambda \mid \Lambda + 2\rho)}{m+g} + \frac{(\Lambda' \mid \Lambda' + 2\rho)}{m'+g} - \frac{\Omega}{m+m'+g} \right) , \tag{10.25}$$

where Ω is the Casimir of $\hat{p}$.

(c) $$[L_k, \hat{p}'] = 0 , \tag{10.26}$$

i.e. the L_k are intertwining operators for the representation of $\hat{p}'$ on $L(\Lambda) \otimes L(\Lambda')$. ∎

LECTURE 11

11.1. $\hat{s\ell}_2$ and its Weyl group.

In this lecture we shall discuss the character formula for a highest weight representation of the simplest affine algebra $\hat{s\ell}_2$. The rather technical results described in this lecture are a necessary preliminary to Lecture 12, where we shall, among other things, give the proof of Lemma 8.6 and thereby complete the proof of the Kac determinant formula.

In the notation of Lecture 9 the Cartan subalgebra $\hat{h}$ of $\hat{s\ell}_2$ is spanned by h_0, h_1 and d where

$$h_0 = c - e_{11}(0) + e_{22}(0) \quad,$$

$$h_1 = e_{11}(0) - e_{22}(0) = \begin{pmatrix} 1 & 0 \\ 0 & -1 \end{pmatrix} .$$

In the following we shall denote the matrix h_1 by α and choose as basis of $\hat{h}$ the elements $\alpha, c = h_0 + h_1$ and d:

$$\hat{h} = \mathbb{C}\alpha \oplus \mathbb{C}c \oplus \mathbb{C}d .$$

The bilinear form $(\cdot | \cdot)$ on $\hat{s\ell}_2$ is nondegenerate when restricted to $\hat{h}$; we see from (9.26) that

$$(\alpha | \alpha) = 2; \quad (c | d) = 1; \quad \text{all other pairs vanish.} \tag{11.1}$$

We shall identify $\hat{h}$ with $\hat{h}^*$ via this form.

In the highest weight representation $L(\Lambda)$ of $\hat{s\ell}_2$ (defined in Lecture 9) the action of $\hat{h}$ on the highest weight vector v_λ is given by

$$h(v_\lambda) = \lambda(h)v_\lambda = (\lambda | h)v_\lambda, \quad h \in \hat{h} \quad .$$

117

The fundamental weights, defined by

$$\omega_i(h_j) = \delta_{ij} \quad (i, j = 0, 1) \; ,$$

are

$$\omega_0 = d, \quad \omega_1 = d + \frac{1}{2}\alpha \; . \tag{11.2}$$

As before, we shall denote by ρ the sum of the fundamental weights:

$$\rho \equiv \omega_0 + \omega_1 = 2d + \frac{1}{2}\alpha \; . \tag{11.3}$$

From (11.2) and Theorem 9.1 we see that $L(\lambda)$ is unitary if and only if λ is of the form

$$\lambda = md + \frac{1}{2}n\alpha + rc, \quad \text{where } r \in \mathbb{R}; \quad m, n \in \mathbb{Z}_+; \quad m \geqslant n \; . \tag{11.4}$$

Then

$$c(v_\lambda) = (\lambda \,|\, c)v_\lambda = mv_\lambda \; , \tag{11.5}$$

and so $L(\lambda)$ is a level m representation.

One of the ingredients which we shall require for the character formula is the Weyl group of $\hat{s\ell}_2$. The Weyl group W of a Lie algebra $\mathfrak{g}$ is the group of those automorphisms of a Cartan subalgebra of $\mathfrak{g}$ which are restrictions of conjugations by elements of G, the Lie group corresponding to $\mathfrak{g}$. In our case this means that the Weyl group is the quotient of the subgroup of $SL_2(\mathbb{C}[t, t^{-1}])$ which leaves $\hat{h}$ invariant under the adjoint action (defined in Lecture 9) by the subgroup which leaves $\hat{h}$ pointwise fixed. (For an equivalent definition, one in terms of reflections of the root system, see Kac [1983].)

The adjoint action of $SL_2(\mathbb{C}[t, t^{-1}])$ on α, c and d can be determined from (9.28). It is then easily checked that the Weyl group of $\hat{s\ell}_2$, which we shall denote by $\hat{W}$, is generated by two elements: one is conjugation by

$$\begin{bmatrix} 0 & 1 \\ -1 & 0 \end{bmatrix}$$

which we denote by r_α, and the second is conjugation by

$$\begin{bmatrix} t & 0 \\ 0 & t^{-1} \end{bmatrix} .$$

We shall denote conjugation by the k-th power of the latter matrix by t_k ($k \in \mathbb{Z}$). Then

$$\hat{W} = \{ t_k , t_k r_\alpha \mid k \in \mathbb{Z} \} , \tag{11.6}$$

since $r_\alpha^2 = 1$ and $t_k r_\alpha = r_\alpha t_{-k}$. The action of the elements of $\hat{W}$ on $\hat{h}$ is easily computed from (9.28):

$$\left. \begin{aligned} r_\alpha(\alpha) &= -\alpha, & r_\alpha(c) &= c, & r_\alpha(d) &= d ; \\ t_k(\alpha) &= \alpha + 2kc, & t_k(c) &= c, & t_k(d) &= d - k\alpha - k^2 c \end{aligned} \right\} \tag{11.7}$$

The relations (11.7) give a matrix representation of r_α and t_k in $\hat{h}$. We shall denote by $\epsilon(w)$ the determinant of $w \in \hat{W}$ in this representation. We find that:

$$\epsilon(r_\alpha) = -1, \quad \epsilon(t_k) = 1, \quad \epsilon(t_k r_\alpha) = -1 . \tag{11.8}$$

11.2. The Weyl-Kac character formula and Jacobi-Riemann theta functions.

Definition 11.1. We define the *character* of the representation $L(\lambda)$ to be the function

$$\text{ch}_\lambda(h) = \text{tr}_{L(\lambda)} \exp(h) \text{ for } h \in \hat{h} .$$

The study of the representations $L(\lambda)$ was started by Kac [1974] with the computation of $\text{ch}_\lambda(h)$ in the framework of arbitrary Kac-Moody algebras. Namely, according to the so-called Weyl-Kac character formula,

$$\text{tr}_{L(\lambda)} \exp(h) = \frac{\sum \epsilon(w)\exp(w(\lambda + \rho) \mid h)}{\sum \epsilon(w)\exp(w(\rho) \mid h)} , \tag{11.9}$$

where the summations run over w in the Weyl group (whenever this series converges). For a proof of (11.9) the reader is referred to Chapter 10 of the book Kac [1983]. We shall now rewrite formula (11.9) in a more explicit form for the case of $\hat{\mathfrak{sl}}_2$ (arbitrary affine algebras can be treated similarly: see Chapter 12 of Kac [1983]).

The calculation of the denominator and numerator in (11.9) being similar, take

$$\mu = md + \frac{n}{2}\alpha + rc .$$
(11.10)

Recalling (11.6) and (11.8), we have:

$$\sum_{w \in \widehat{W}} \epsilon(w)\exp(w(\mu) \mid h) = \sum_{k \in \mathbb{Z}} \exp(t_k(\mu) \mid h) - \sum_{k \in \mathbb{Z}} \exp(t_k r_\alpha(\mu) \mid h) .$$

From (11.7) and (11.10), for each $k \in \mathbb{Z}$ we have:

$$t_k(\mu) = md + (\frac{1}{2} n - mk)\alpha + (r + kn - k^2 m)c ,$$
(11.11a)

$$t_k r_\alpha(\mu) = md - (\frac{1}{2} n + mk)\alpha + (r - kn - k^2 m)c .$$
(11.11b)

We choose h to have the general form

$$h = 2\pi i(\frac{1}{2} z\alpha - \tau d + uc), \quad \text{where } \tau, z, u \in \mathbb{C} .$$
(11.12)

Then from (11.11a, b) and (11.1), for each $k \in \mathbb{Z}$ we have:

$$(t_k(\mu) \mid h) = 2\pi i[(\frac{1}{2} n - mk)z - (r + kn - k^2 m)\tau + mu] ,$$
(11.13a)

$$(t_k r_\alpha(\mu) \mid h) = 2\pi i[-(\frac{1}{2} n + mk)z - (r - kn - k^2 m)\tau + mu] .$$
(11.13b)

Replace k in (11.13a) by $n/2m - k$ (so that now $k \in n/2m + \mathbb{Z}$) and in (11.13b) by $-n/2m - k$ (so that $k \in -n/2m + \mathbb{Z}$); the right-hand sides of (11.13a, b) become respectively:

$$2\pi i(mkz + (mk^2 - (r + \frac{n^2}{4m}))\tau + mu) \quad (k \in \frac{n}{2m} + \mathbb{Z}) ,$$
(11.14a)

$$2\pi i(mkz + (mk^2 - (r + \frac{n^2}{4m}))\tau + mu) \quad (k \in -\frac{n}{2m} + \mathbb{Z}) .$$
(11.14b)

Definition 11.2. We define $\Theta_{n,m}(\tau, z, u)$ by

$$\Theta_{n,m}(\tau, z, u) = \exp(2\pi imu) \sum_{k \in \frac{n}{2m} + \mathbb{Z}} \exp(2\pi im(k^2\tau + kz)) .$$
(11.15)

Remark 11.1. These are the celebrated Jacobia-Riemann theta functions. Clearly $\Theta_{n,m}(\tau, z, u)$ converges absolutely for τ in the upper half complex plane and arbitrary complex z and u.

We thus obtain

$$\sum_{w \in \widehat{W}} \epsilon(w)\exp(w(\mu) \,|\, h) = q^{-(r + \frac{n^2}{4m})} (\Theta_{n,m}(\tau, z, u) - \Theta_{-n,m}(\tau, z, u)) .$$
(11.16)

Here and further, q stands for exp $2\pi i\tau$.

In the numerator of (11.9) we have $\mu = \lambda + \rho = (m + 2)d + \frac{1}{2}(n + 1)\alpha + rc$, while in the denominator we have $\mu = \rho = 2d + \frac{1}{2}\alpha$. Substituting in (11.6) we arrive at:

Proposition 11.1. For $\lambda = md + \frac{1}{2} n\alpha + rc$ as in (11.4) and $z, \tau, u \in \mathbb{C}$, *Im* $\tau > 0$, one has:

$$\text{ch}_\lambda(h) \equiv \text{tr}_{L(\lambda)} \exp[\, 2\pi i(\frac{1}{2} z\alpha - \tau d + uc)]$$

$$= q^{-s_\lambda} \frac{\Theta_{n+1,m+2}(\tau, z, u) - \Theta_{-n-1,m+2}(\tau, z, u)}{\Theta_{1,2}(\tau, z, u) - \Theta_{-1,2}(\tau, z, u)} ,$$
(11.17a)

where

$$s_\lambda = \frac{(n + 1)^2}{4(m + 2)} - \frac{1}{8} + r .$$
(11.17b)

∎

Remark 11.2. Dividing the numerator and the denominator of (11.17a) by $1 - e^{-2\pi iz}$, one derives from (11.7a), as $u = 0$ and $z \to 0$, a formula for the q-dimension (= partition function):

$$\text{tr}_{L(md+\frac{1}{2} n\alpha)} \, q^{-d} \; = \; \varphi(q)^{-3} \sum_{j \in \mathbb{Z}} \; (2j(m+2)+n+1)q^{(m+2)j^2+(n+1)j} \; .$$

The theta functions $\Theta_{n,m}(\tau,z,u)$ have an important multiplication rule which we shall require (Kac-Peterson [1984a]):

Proposition 11.2. $\Theta_{n,m}(\tau,z,u) \, \Theta_{n',m'}(\tau,z,u)$

$$= \sum_j d_j^{(m,m',n,n')}(q) \, \Theta_{n+n'+2mj, \, m+m'}(\tau,z,u) \quad (j \in \mathbb{Z} \; \text{mod}(m+m')\mathbb{Z}) \, ,$$

where

$$d_j^{(m,m',n,n')}(q) = \sum_k q^{mm'(m+m')k^2} \quad (k \in \mathbb{Z} + \frac{m'n - mn' + 2jmm'}{2mm'(m+m')}) \; .$$

Proof.   Without loss of generality we can put $u=0$. Then

$$\Theta_{n,m} = \sum_k q^{mk^2} \exp(2\pi imkz) \quad (k \in \frac{n}{2m} + \mathbb{Z}) \, ,$$

$$\Theta_{n',m'} = \sum_{k'} q^{m'k'^2} \exp(2\pi im'k'z) \quad (k' \in \frac{n'}{2m'} + \mathbb{Z}) \, ,$$

and

$$\Theta_{n,m} \, \Theta_{n',m'} = \sum_{k,k'} q^{mk^2+m'k'^2} \exp[\, 2\pi i(mk + m'k')z \,] \; .$$

Let $k = (n/2m) + i$, $k' = (n'/2m') + i'$, where $i, i' \in \mathbb{Z}$. Define s, s' by

$$(m+m')s = k - k' = \frac{nm' - n'm}{2mm'} + i - i' \, ,$$

$$(m+m')s' = mk + m'k' = (m+m')(k' + ms) \; .$$

Let $i - i' = (m+m')\ell + j$ where $\ell \in \mathbb{Z}$ and $j \in \mathbb{Z} \; \text{mod}(m+m')\mathbb{Z}$. Then

$$s \in \frac{nm' - n'm + 2mm'j}{2mm'(m+m')} + \mathbb{Z} \, , \quad s' \in \frac{n + n' + 2mj}{2(m+m')} + \mathbb{Z} \; .$$

This gives us a bijection between pairs (k, k') and triples (s, s', j). Noting that

$$mk^2 + m'k'^2 = mm'(m + m')s^2 + (m + m')s'^2 ,$$

we get:

$$\Theta_{n,m} \, \Theta_{n',m'} = \sum_j \, (\sum_s q^{mm'(m+m')s^2}) (\sum_{s'} q^{(m+m')s'^2} \exp 2\pi i(m + m')s'z)$$

which proves the proposition. ∎

We shall now consider the character formula (11.17a, b) for the simplest nontrivial case of $\lambda = d$ $(m = 1, n = 0, r = 0)$. We call this the *basic representation*. Since $s_d = -1/24$, we get

$$\text{ch}_d(h) = q^{1/24} \frac{\Theta_{1,3} - \Theta_{-1,3}}{\Theta_{1,2} - \Theta_{-1,2}} .$$

We shall now use Proposition 11.2 to prove a much simpler formula for $\text{ch}_d(h)$:

Proposition 11.3. $$\text{ch}_d(h) = \frac{\Theta_{0,1}}{\varphi(q)} , \qquad (11.18)$$

where $\varphi(q)$ is the function defined in (2.8b).

Proof. We have to prove that

$$\Theta_{0,1}(\Theta_{1,2} - \Theta_{-1,2}) = q^{1/24} \varphi(q) (\Theta_{1,3} - \Theta_{-1,3}) .$$

Applying Proposition 11.2 to the left-hand side we obtain

$$(\Theta_{1,3} - \Theta_{-1,3}) \left(\sum_{k \in -\frac{1}{12} + \mathbb{Z}} q^{6k^2} - \sum_{k' \in \frac{5}{12} + \mathbb{Z}} q^{6k'^2} \right)$$

$$= (\Theta_{1,3} - \Theta_{-1,3}) q^{1/24} \sum_{k \in \mathbb{Z}} (-1)^j q^{(3j^2+j)/2} .$$

A famous identity due to Euler tells us that

$$\varphi(q) = \sum_{k \in \mathbb{Z}} (-1)^k q^{(3k^2+k)/2} \qquad (11.19)$$

which completes the proof.   ∎

Remark 11.3. The Weyl-Kac character formula (11.9) is often written in a different form:

$$\text{tr}_{L(\lambda)} \exp(h) = \frac{\sum\limits_{w} \epsilon(w) \exp[\,(w(\lambda+\rho)\,|\,h) - (\rho\,|\,h)\,]}{\prod\limits_{\substack{\gamma\in\Delta \\ (\rho|\gamma)>0}} (1 - \exp[-(\gamma\,|\,h)])^{\text{mult}\,\gamma}} \quad . \tag{11.20}$$

Here Δ is the set of roots and mult γ is the multiplicity of $\gamma \in \Delta$. (Recall that $\gamma \in \hat{h}$ is called a root if the equation $[h, x] = (\gamma\,|\,h)x$ has a non-zero solution for all $h \in \hat{h}$; the dimension of the space of solutions is called the multiplicity of γ). Putting $\lambda = 0$ in (11.20) we obtain the so called Weyl-Macdonald-Kac formula:

$$\prod_{\substack{\gamma\in\Delta \\ (\rho|\gamma)>0}} (1 - \exp[-(\gamma\,|\,h)])^{\text{mult}\,\gamma} = \sum_{w} \epsilon(w) \exp[(w(\rho)\,|\,h) - (\rho\,|\,h)] \quad . \tag{11.21}$$

Plugging (11.21) into (11.20) gives (11.9). In the case of $\hat{s\ell}_2$, we have $\Delta = \{\pm\alpha + kd, kd \mid k \in \mathbb{Z}\}$ and the multiplicities of all roots are 1. Putting $u = \exp[-(h_0\,|\,h)]$, $v = \exp[-(h_1\,|\,h)]$, it is easy to see that (11.21) turns into the classical Jacobi triple product identity:

$$\prod_{k\geqslant 1} (1 - u^{k-1}v^k)(1 - u^k v^{k-1})(1 - u^k v^k) = \sum_{j\in\mathbb{Z}} (-1)^j u^{\frac{j(j+1)}{2}} v^{\frac{j(j-1)}{2}} \quad .$$

Putting $u = q$, $v = q^2$ in this identity gives the Euler identity (11.19). See Chapter 12 of Kac [1983] for more details.

11.3 A character identity.

The following character identity will play a crucial role in Lecture 12.

Proposition 11.4. For $\lambda = md + \frac{1}{2}n\alpha$ $(m \geqslant n \geqslant 0)$, one has

$$\text{ch}_d \text{ch}_\lambda = \sum_{k\in I} \psi_{m,n;k}\,\text{ch}_{d+\lambda-k\alpha} \quad , \tag{11.22a}$$

where

$$I = \left\{ k \in \mathbb{Z} \,\left|\, -\frac{1}{2}(m + 1 - n) \leqslant k \leqslant \frac{1}{2}n \right. \right\} \quad , \tag{11.22b}$$

and

$$\psi_{m,n;k} = (f_k^{(m,n)} - f_{n+1-k}^{(m,n)})/\varphi(q) , \qquad (11.22c)$$

where

$$f_k^{(m,n)} = \sum_{j \in \mathbb{Z}} q^{(m+2)(m+3)j^2 + ((n+1)+2k(m+2))j + k^2} . \qquad (11.22d)$$

Proof. From (11.17) and (11.18),

$$\varphi(q)\, \text{ch}_d \text{ch}_\lambda = \frac{\Theta_{0,1}\Theta_{n+1,m+2} - \Theta_{0,1}\Theta_{-n-1,m+2}}{\Theta_{1,2} - \Theta_{-1,2}} q^{-\frac{(n+1)^2}{4(m+2)} + \frac{1}{8}} .$$

By Proposition 11.2:

$$\Theta_{0,1}\Theta_{n+1,m+2} = \sum_\ell a_\ell \, \Theta_{n+1,m+2} \quad (\ell \in \mathbb{Z} \bmod (m+3)\mathbb{Z}) ,$$

with

$$a_\ell = \sum_i q^{(m+2)(m+3)i^2} ,$$

where $i = j + [-(n+1) + 2\ell(m+2)]/[2(m+2)(m+3)]$ and $j \in \mathbb{Z}$. Note that

$$(m+2)(m+3)i^2 - \frac{(n+1)^2}{4(m+2)} = (m+2)(m+3)j^2$$

$$+ (-(n+1) + 2\ell(m+2))j + \ell^2 - \frac{(n+1+2\ell)^2}{4(m+3)} .$$

Hence,

$$\Theta_{0,1}\Theta_{n+1,m+2} \, q^{-\frac{(n+1)^2}{4(m+2)}} = \sum_\ell b_\ell \, \Theta_{n+1+2\ell,m+3} \, q^{-\frac{(n+1+2\ell)^2}{4(m+3)}} ,$$

where

$$b_\ell = \prod_{j \in \mathbb{Z}} q^{(m+2)(m+3)j^2 + (-(n+1)+2\ell(m+2))j + \ell^2} .$$

Putting $\ell = -k$ and making the transformation $j \to -j$, we see that $b_{-k} = f_k^{(m,n)}$, so that

$$\Theta_{0,1} \Theta_{n+1,n+2} \; q^{-\frac{(n+1)^2}{4(m+2)}} = \sum_k f_k^{(m,n)} \, \Theta_{n+1-2k,m+3} \; q^{-\frac{(n+1-2k)^2}{4(m+3)}} \; ,$$

where $k \in \mathbb{Z} \bmod (m+3)\mathbb{Z}$. Similarly,

$$\Theta_{0,1} \Theta_{-n-1,m+2} \; q^{-\frac{(n+1)^2}{4(m+2)}} = \sum_k f_k^{(m,n)} \, \Theta_{-n-1+2k,\,m+3} \; q^{-\frac{(n+1-2k)^2}{4(m+3)}}$$

where $k \in \mathbb{Z} \bmod (m+3)\mathbb{Z}$.

Hence

$$\mathrm{ch}_d \, \mathrm{ch}_\lambda = \sum_k \frac{f_k^{(m,n)}}{\varphi(q)} \; \mathrm{ch}_{d+\lambda-k\alpha} \qquad (k \in \mathbb{Z} \bmod (m+3)\mathbb{Z}) \; . \tag{11.23}$$

Now $d + \lambda - k\alpha = (m+1)d + \frac{1}{2}(n-2k)\alpha$; this corresponds to a non-negative integral linear combination of the fundamental weights ω_0 and ω_1, if and only if $k \in I$, where I was defined in (11.22b). We use the freedom in choosing the domain of definition of k in (11.23), i.e. of $\mathbb{Z} \bmod (m+3)\mathbb{Z}$, by taking k to be in the union of the sets I, J and K, where:

$$I = \left\{ k \in \mathbb{Z} \; \middle| \; -\frac{1}{2}(m+1-n) \leqslant k \leqslant \frac{1}{2}n \right\} \; ,$$

$$J = \left\{ k \in \mathbb{Z} \; \middle| \; \frac{1}{2}n + 1 \leqslant k \leqslant \frac{1}{2}(m+3+n) \right\} \; ,$$

$$K = \left\{ k \in \mathbb{Z} \; \middle| \; k = \frac{n+1}{2}, \quad k = \frac{m+n}{2} + 2 \right\} \; .$$

Note that K is nonempty only if n is odd, or if $m+n$ is even. The following symmetry property of the function $\mathrm{ch}_{d+\lambda-k\alpha}$ is easily checked:

$$\text{if } k \to n + 1 - k, \quad \text{then } \mathrm{ch}_{d+\lambda-k\alpha} \to - \, \mathrm{ch}_{d+\lambda-k\alpha} \; . \tag{11.24}$$

An immediate corollary is that,

$$\text{ch}_{d+\lambda-k\alpha} = 0 \quad \text{if } 2k = n + 1 \, (\text{mod } m + 3) \, .$$

Hence, $\text{ch}_{d+\lambda-k\alpha} = 0$ for $k \in K$. Then, since J maps onto I under the transformation $k \to n + 1 - k$, (11.23) becomes (11.22) on using (11.24). ∎

LECTURE 12

12.1. Preliminaries on $\hat{s\ell}_2$.

In this lecture we shall use the Goddard-Kent-Olive (GKO) construction, described in Lecture 10, to construct unitary highest weight representations $V(c, h)$ of the Virasoro algebra for all pairs (c, h) given by the discrete series:

$$c = c_m \equiv 1 - \frac{6}{(m + 2)(m + 3)} \qquad (m = 0, 1, 2 \ldots) , \qquad (12.1a)$$

$$h = h_{r,s}^{(m)} \equiv \frac{[(m + 3)r - (m + 2)s]^2 - 1}{4(m + 2)(m + 3)} \qquad (r, s \in \mathbb{N}, \ 1 \leqslant s \leqslant r \leqslant m + 1) . \qquad (12.1b)$$

In the course of the construction we shall also be able to prove Lemma 8.6 and thereby complete the proof of the Kac determinant formula. The unitarity of the discrete series was proved independently by Goddard-Kent-Olive [1986], Kac-Wakimoto [1986] and Tsuchiya-Kanie [1986]. This lecture follows closely the paper Kac-Wakimoto [1986].

We now summarize some properties of $\hat{s\ell}_2$ and $\hat{s\ell}_2$ that we shall use. We choose the standard basis of $\hat{s\ell}_2$: $u_1 = e, u_2 = \alpha, u_3 = f$, where

$$e = \begin{bmatrix} 0 & 1 \\ 0 & 0 \end{bmatrix}, \qquad \alpha = \begin{bmatrix} 1 & 0 \\ 0 & -1 \end{bmatrix}, \qquad f = \begin{bmatrix} 0 & 0 \\ 1 & 0 \end{bmatrix} .$$

Then

$$\hat{n}_+ = \mathbb{C}e + \sum_{k > 0} t^k \hat{s\ell}_2 .$$

Defining the dual basis $\{u^i \mid i = 1, 2, 3\}$ by (10.2) with respect to the trace form $(a \mid b) = \operatorname{tr} ab$, we have

$$u^1 = f, \quad u^2 = \alpha/2, \quad u^3 = e .$$

Then, by Lemma 10.1, the Casimir operator Ω_0 is given by

$$\Omega_0 = u_i u^i = ef + fe + \alpha^2/2 = \alpha^2/2 + \alpha + 2fe \ . \qquad (12.2)$$

(Note that in (12.2) we have used the commutation relations to move the element e to the right.) We can compute the eigenvalue of the Casimir in its adjoint representation directly and we find that

$$g = 2 \ , \qquad (12.3)$$

which is in agreement with Table 10.1.

Let

$$P_+^0 = \left\{ md + \frac{1}{2} n\alpha \ \bigg| \ m, n \in \mathbf{Z}_+, \ m \geq n \right\}; \qquad P_+ = P_+^0 + \mathbf{R}c \ . (12.4)$$

Recall from Lecture 9 that given $\lambda = md + \frac{1}{2}n\alpha + rc \in P_+$ there exists a unique (up to equivalence) irreducible unitary highest weight representation of $\hat{s\ell}_2$ on a complex vector space $L(\lambda)$. This representation of $\hat{s\ell}_2$ remains irreducible and is independent of r when viewed as a representation of $\hat{s\ell}_2'$.

12.2. A tensor product decomposition of some representation of $\hat{s\ell}_2$.

We shall now consider the tensor product of two unitarizable highest weight representation of $\hat{s\ell}_2$, *viz.* $L(d)$ and $L(\lambda)$ where

$$\lambda = md + \frac{1}{2} n\alpha \in P_+^0 \ . \qquad (12.5)$$

The GKO construction gives us a unitary representation of *Vir* which commutes with $\hat{s\ell}_2'$. Hence the space $L(d) \otimes L(\lambda)$ can be reduced with respect to the direct sum of *Vir* and $\hat{s\ell}_2'$.

An essential tool in the following will be the character of the representations of $\hat{s\ell}_2$, which is computed from the Weyl-Kac character formula as we saw in Lecture 11. The usefulness of the character is due to its simple algebraic properties: the character of a tensor product of representations is the product of their characters while the character of a direct sum is the sum of their characters. Moreover two representations are equivalent if and only if their characters are equal.

The characters of $L(d)$ and $L(\lambda)$ were denoted ch_d and ch_λ in Lecture 11. The character of $L(d) \otimes L(\lambda)$ is therefore $\mathrm{ch}_d \, \mathrm{ch}_\lambda$. We derived an important identity for this quantity in Proposition 11.4. We shall, however, have to do some more work to bring it to the form in which we need it:

Lemma 12.1. Let

$$r = n + 1, \qquad s = n + 1 - 2k \qquad \text{if } k \geqslant 0$$
$$r = m - n + 1, \quad s = m - n + 2 + 2k \quad \text{if } k < 0$$
(12.6)

(Note that $1 \leqslant s \leqslant r \leqslant m + 1$ since $m \geqslant n$ and $m, n \in \mathbb{Z}_+$.) Then

$$\varphi(q) q^{-k^2} \psi_{m,n;k} \equiv q^{-k^2} (f_k^{(m,n)} - f_{n+1-k}^{(m,n)}) = A + B + C \text{ ,} \quad (12.7a)$$

where

$$A = 1 - q^{rs} - q^{(m+2-r)(m-3-s)} \text{ ,}$$

$$B = \sum_{j \in \mathbb{N}} q^{(m+2)(m+3)j^2 + ((m+3)r - (m+2)s)j} (1 - q^{2(m+2)sj+rs}) \text{ ,}$$
(12.7b)

$$C = \sum_{j \in \mathbb{N}} q^{(m+2)(m+3)j^2 - ((m+3)r - (m+2)s)j} (1 - q^{2(m+2)(m+3-s)j + (m+2-r)(m+3-s)}).$$

Proof. A straightforward calculation using the definition (11.22c, d). ∎

Remark 12.1. Note from (12.7) that

$$\psi_{m,n;k} = \frac{q^{k^2}}{\varphi(q)} (1 - q^{rs} - q^{(m+2-r)(m+3-s)} + \text{ higher powers of } q) \quad (12.8)$$

Proposition 12.1. Let $\lambda = md + \frac{1}{2} n\alpha$ ($m, n \in \mathbb{Z}_+$; $m \geqslant n$). Then,

$$\text{ch}_d \, \text{ch}_\lambda = \sum_{k \in I} \sum_{j \in \mathbb{Z}_+} \Delta_{m,n;k}^j \, \text{ch}_{d+\lambda-k\alpha-jc} \quad (12.9a)$$

where

$$I = \left\{ k \in \mathbb{Z} \; \middle| \; -\frac{1}{2} (m + 1 - n) \leqslant k \leqslant \frac{1}{2} n \right\} \quad (12.9b)$$

The $\Delta_{m,n;k}^j$ are non-negative integers defined by the expansion

$$\psi_{m,n;k} = \sum_{j \in \mathbb{Z}_+} \Delta_{m,n;k}^j \, q^j \text{ .} \quad (12.10)$$

The minimum value of j appearing with a nonzero coefficient in (12.10) is $j = k^2$, and $\Delta_{m,n;k}^{k^2} = 1$.

Proof. The fact that $\Delta_{m,n;k}^{j} \in \mathbb{Z}_{+}$ is clear from their representation-theoretical meaning explained below.

To prove the last statement of the proposition, if suffices to show that A, B and C in (12.7b) have each an expansion in powers of q with nonnegative integer coefficients when multiplied by $\varphi(q)^{-1}$. Note that in B and C the term in brackets is always of the form $1 - q^i (i > 0)$. Hence on multiplying by $\varphi(q)^{-1} = \prod_{j=1}^{\infty} (1 - q^j)^{-1}$ one such factor cancels. The remainder will have an expansion of the required kind. It is straightforward to use the expansion (2.8a) to show by direct multiplication that $A/\varphi(q)$ has the required expansion.

Noting that the lowest power of q in (12.8) is q^{k^2}, this proves the last statement of the proposition. From (11.17) we see that

$$\mathrm{ch}_{\lambda - jc} = q^j \, \mathrm{ch}_{\lambda} \ .$$

Recalling Proposition 11.4, this completes the proof of Proposition 12.2. ∎

Proposition 12.1 has a very simple representation-theoretical meaning. Equation (12.9) expresses the decomposition of $L(d) \otimes L(\lambda)$ into a direct sum of unitary $\widehat{s\ell}_2$ representations $L(d + \lambda - k\alpha - jc)$:

$$L(d) \otimes L(\lambda) = \sum_{k \in I} \sum_{j \in \mathbb{Z}_+} \Delta_{m,n;k}^{j} L(d + \lambda - k\alpha - jc) \ , \quad (12.11)$$

and $\Delta_{m,n;k}^{j}$ is simply the multiplicity of the occurence of $L(d + \lambda - k\alpha - jc)$ in this decomposition.

12.3. Construction and unitarity of the discrete series representations of *Vir*.

Let $U_{m,n;k}^{(j)}$ denote the space of highest weight vectors of $\widehat{s\ell}_2$ in $L(d) \otimes L(\lambda)$ with highest weight $d + \lambda - k\alpha - jc \in P_+$. Then

$$\Delta_{m,n;k}^{j} = \dim U_{m,n;k}^{(j)} \ . \quad (12.12)$$

Clearly $U_{m,n;k}^{(j)}$ is an eigenspace of d with eigenvalue $(d \mid d + \lambda - k\alpha - jc) = -j$. Now define

$$U_{m,n;k} = \bigoplus_{j \in \mathbb{Z}_+} U_{m,n;k}^{(j)} \ . \quad (12.13)$$

It is clear that $U_{m,n;k}$ is the space of highest weight vectors of $\hat{s\ell}'_2$ in $L(d) \otimes L(\lambda)$ with highest weight $d + \lambda - k\alpha$. Now the GKO construction produces a representation of *Vir* which commutes with $\hat{s\ell}'_2$ and hence *Vir* maps $U_{m,n;k}$ into itself. The central charge of this representation of *Vir* in $U_{m,n;k}$ is given by (10.24) with $m' = 1, g = 2$ and dim $p = 3$, which gives $c = c_m$ $(m \in \mathbb{Z}_+)$, where c_m is given by (12.1a). The operator L_0 in this representation is given by (10.25) with

$$\Lambda = d, \quad \Lambda' = md + \frac{1}{2} n\alpha, \quad 2\rho = 4d + \alpha .$$

Thus

$$L_0 = \frac{n(n+2)}{4(m+2)} - \frac{\Omega}{2(m+3)} .$$

From (10.18) and (10.19),

$$\Omega = 2(c+2)d + \Omega_0 + 2 \sum_{j>0} u_i(-j)u^i(j) ,$$

where $\Omega_0 = \alpha^2/2 + \alpha + 2fe$ from (12.2). Each $v \in U_{m,n;k}$ is a highest weight vector and hence is annihilated by the $u(j)$ with $j > 0$ and by e. Thus

$$\Omega(v) = (2d(c+2) + \alpha^2/2 + \alpha)(v) \quad \text{for } v \in U_{m,n;k} .$$

Since

$$d + \lambda - k\alpha = (m+1)d + \frac{n-2k}{2} \alpha = \Lambda \quad \text{(say)} ,$$

$$(c + 2 \mid \Lambda) = m + 3, \quad (\alpha \mid \Lambda) = n - 2k ,$$

we obtain:

$$\Omega(v) = (2d(m+3) + \frac{(n-2k)^2}{2} + (n-2k))v \quad \text{for } v \in U_{m,n;k} .$$

Thus in $U_{m,n;k}$ we have

$$L_0 = -d + \frac{n(n+2)}{4(m+2)} - \frac{(n-2k)(n-2k+2)}{4(m+3)} . \qquad (12.14)$$

To determine the minimal eigenvalue of L_0 on $U_{m,n;k}$ we need to know the minimal eigenvalue of $(-d)$ on $U_{m,n;k}$. According to (12.13), $U_{m,n;k}$ is the direct sum of eigenspaces of $(-d)$: $U_{m,n;k}^{(j)}$ is an eigenspace of $(-d)$ with eigenvalue j. Hence we need to know the minimal value of j for which $\Delta_{m,n;k}^j$ is nonzero (see (12.12)). This was determined in Proposition 12.1 to be $j = k^2$. Thus the minimal eigenvalue of L_0 in the representation of Vir on $U_{m,n;k}$ is

$$h = k^2 + \frac{n(n+2)}{4(m+2)} - \frac{(n-2k)(n-2k+2)}{4(m+3)} . \qquad (12.15)$$

Changing from the variables m, n, k to m, r, s by (12.6), we find that $h = h_{r,s}^{(m)}$ as given by (12.1b).

Denoting $U_{m,n;k}$ in the following by $U_{r,s}^{(m)}$, we have thus constructed a unitary representation of Vir on the space $U_{r,s}^{(m)}$ such that the central charge is c and the minimal eigenvalue of L_0 is h for every pair (c, h) in the discrete series (12.1a, b). All the eigenvalues of L_0 are from $h + \mathbb{Z}_+$. The highest component of the representation of Vir on $U_{r,s}^{(m)}$, i.e. the subrepresentation generated by the eigenvector of L_0 with the minimal eigenvalue $h = h_{r,s}^{(m)}$, is an irreducible unitary highest weight representation. This completes the proof that all representations $V(c, h)$ with (c, h) from the list (12.1a, b) are unitary.

From the above discussion we also obtain:

$$\text{ch } V(c_m, h_{r,s}^{(m)}) \leqslant \text{ch } U_{r,s}^{(m)}. \qquad (12.16)$$

From (12.14) and (12.15) we see that $U_{m,n;k}^{(j)}$ is an eigenspace of L_0 with eigenvalue $(j - k^2) + h_{r,s}^{(m)}$. From (12.10) and (12.12) we have

$$\psi_{m,n;k} = \sum_{j \in \mathbb{Z}_+} \dim U_{m,n;k}^{(j)} \, q^j .$$

Thus $\psi_{m,n;k}$ is essentially the character of this representation of Vir. More precisely,

$$\text{ch } U_{m,n;k} \;=\; q^{h_{r,s}^{(m)}}\, q^{-k^2}\, \psi_{m,n;k} \;\;,$$

where $\psi_{m,n;k}$ is given by (12.7a, b). From (12.16) and Remark 12.1 we have:

$$\text{ch } V(c_m,\, h_{r,s}^{(m)}) \;\leqslant\; \frac{q^{h_{r,s}^{(m)}}}{\varphi(q)}\, (1 \,-\, q^{rs} \,-\, q^{(m+2-r)(m+3-s)}$$

$$+ \text{ higher degree terms in } q) \; . \tag{12.17}$$

Remark 12.2. We can now interpret Proposition 11.4 as saying that for $\lambda = md + \tfrac{1}{2}\, n\alpha \in P_+^0$ we have with respect to $\hat{s\ell}_2' \oplus Vir$:

$$L(d) \otimes L(md + \tfrac{1}{2}\, n\alpha) \;=\; \bigoplus_{\substack{0 \leqslant j \leqslant n \\ j \equiv n \,(\mathrm{mod}\,2)}} L((m+1)d + \tfrac{1}{2}\, j\alpha) \oplus U_{n+1,\,j+1}^{(m)}$$

$$\bigoplus_{\substack{n+1 \leqslant j \leqslant m+1 \\ j \equiv n \,(\mathrm{mod}\,2)}} L((m+1)d + \tfrac{1}{2}\, j\alpha) \otimes U_{m-n+1,\,m+2-j}^{(m)} \; . \tag{12.18}$$

Keeping m fixed in (12.18) and varying n in the range $0 \leqslant n \leqslant m$, we see that on the right-hand side we get all $U_{r,s}^{(m)}$ satisfying $1 \leqslant s \leqslant r \leqslant m+1$, each occuring exactly once. It follows that with respect to Vir:

$$\Big[L(d) \otimes \sum_{\substack{n \in \mathbb{Z}_+ \\ 0 \leqslant n \leqslant m}} L(md + \tfrac{1}{2}\, n\alpha) \Big]^{\hat{n}_+} \;=\; \bigoplus_{\substack{r,s \in \mathbb{N} \\ 1 \leqslant s \leqslant r \leqslant m+1}} U_{r,s}^{(m)} \;\;, \tag{12.19}$$

where the notation on the left-hand side means the subspace in the tensor product annihilated by $\hat{n}_+$. If we extend the sums in (12.19) over all possible values of $m \in \mathbb{Z}_+$, we see that every representation of the discrete series appears at least once in this space.

We summarize below the results obtained:

Theorem 12.1. (a) The irreducible representation $V(c, h)$ of Vir is unitary for $c = c_m$, $h = h_{r,s}^{(m)}$ (given by (12.1a, b)), where $m, r, s \in \mathbb{Z}_+$, $1 \leqslant s \leqslant r \leqslant m+1$, and for $c \geqslant 1, h \geqslant 0$.

(b) With respect to *Vir* we have:

$$\left[L(d) \otimes \sum_{\substack{m,n \in \mathbb{Z}_+ \\ m \geqslant n}} L\left(md + \frac{1}{2}n\alpha\right) \right]^{\widehat{n}_+} = \bigoplus_{\substack{m,r,s \in \mathbb{Z}_+ \\ 1 \leqslant s \leqslant r \leqslant m+1}} U_{r,s}^{(m)} \; ,$$

where the highest component of $U_{r,s}^{(m)}$ is $V(c_m, h_{r,s}^{(m)})$.

(c) $\operatorname{ch} V(c_m, h_{r,s}^{(m)}) \leqslant \operatorname{ch} U_{r,s}^{(m)}$

$$= \frac{q^{h_{r,s}^{(m)}}}{\varphi(q)} \left(1 - q^{rs} - q^{(m+2-r)(m+3-s)} + B + C\right) , \tag{12.20a}$$

where B and C are given by (12.7b).
Equivalently:

$$q^{-c_m/24} \operatorname{ch} V(c_m, h_{r,s}^{(m)}) \leqslant q^{-c_m/24} \operatorname{ch} U_{r,s}^{(m)}$$

$$= \frac{1}{\eta(\tau)} \left(\Theta_{r(m+3)-s(m+2),\,(m+2)(m+3)}(\tau, 0, 0)\right.$$

$$\left. - \Theta_{r(m+3)+s(m+2),\,(m+2)(m+3)}(\tau, 0, 0)\right) , \tag{12.20b}$$

where $$\eta(\tau) = q^{1/24}\,\varphi(q) . \tag{12.20c}$$

(Here $1 \leqslant s \leqslant r \leqslant m+1, m, r, s \in \mathbb{Z}_+$.)

Remark 12.3. In fact, in (12.16), (12.17) and (12.20a, b) one has equality; in other words the representation of *Vir* on $U_{r,s}^{(m)}$ is irreducible and coincides with $V(c_m, h_{r,s}^{(m)})$. The most straightforward way of proving this is to calculate $\operatorname{ch} V(c_m, h_{r,s}^{(m)})$ and to notice that it coincides with the right-hand side of (12.20). The computation of $\operatorname{ch} V(c, h)$ consists of two steps, both of which are based on the Kac determinant formula. First, one finds all possible inclusions of Verma representations and the irreducible subfactors contained in $M(c, h)$; this was established by Kac [1978a] using the Jantzen filtration (see e.g. Kac-Kazhdan [1979]). Secondly, one shows that the irreducible subfactors of $M(c, h)$ occur with multiplicity one. This is the more difficult part, conjectured by Kac [1982] and proved by Feigin and Fuchs [1983a, b], [1984]. The explicit formulae for all ch $V(c, h)$ are easily derived from these facts: see Feigin-Fuchs [1983b], [1984]. Thus Theorem 12.1(b) provides a "model" for the discrete series of *Vir*, i.e. the space in which all the representations of the discrete series appear and exactly once.

Remark 12.4. Formula (12.20b) shows that the functions $q^{-c_m/24}$ ch $V(c_m, h_{r,s}^{(m)})$ are modular functions in $\tau (Im \, \tau > 0)$, where as before $q = \exp 2\pi i\tau$. Also, Remark 11.2 shows that the same is true for the functions $q^{-s_\lambda} \text{tr}_{L(\lambda)} q^{-d}$. This observation has been playing an important role in recent developments in representation theory, as well as in the theory of 2-dimensional statistical models and string theory (see Kac-Wakimoto [1987], Capelli-Itzykson-Zuber [1987], Gepner-Witten [1986] and references there).

12.4. Completion of the proof of the Kac determinant formula.

We shall now show how the inequality for ch $V(c_m, h_{r,s}^{(m)})$ given by (12.17) can be used to prove Lemma 8.6, thereby completing the proof of the Kac determinant formula discussed in Lecture 8. We recall from (3.15) that the character of the Verma representation $M(c_m, h_{r,s}^{(m)})$ is $q^{h_{r,s}^{(m)}}/\varphi(q)$. Let $r' = m + 2 - r$, $s' = m + 3 - s$ so that $s' > r'$. Note from (12.1b) that $h_{r,s}^{(m)} = h_{r',s'}^{(m)}$. From the occurence in (12.17) of the lowest powers of q, viz. q^{rs} and $q^{r's'}$, with negative coefficients, we can deduce that $J(c_m, h_{r,s}^{(m)})$ (the maximal proper subrepresentation of $M(c_m, h_{r,s}^{(m)})$) has a nonzero component at each level $n \geqslant \min (rs, r's')$. Thus $\det_n (c, h)$ has a zero at $h = h_{r,s}^{(m)}$ $(1 \leqslant s \leqslant r \leqslant m + 1)$ for $n \geqslant \min (rs, r's')$. Thus $\det_n (c_m, h)$ has a zero at $h = h_{r,s}^{(m)}$ for all pairs (r, s) satisfying $1 \leqslant r$, $s \leqslant m + 1$ and $n \geqslant rs$. In Theorem 8.1 we defined $\varphi_{r,s}(c, h) = (h - h_{r,s})(h - h_{s,r})$ for $r \neq s$ and $\varphi_{r,r}(c, h) = h - h_{r,r}$. Viewed as a polynomial in the two variables c, h, it is clear that $\varphi_{r,s}(c, h)$ is irreducible (over the complex field), i.e. it cannot be written as the product of linear factors (for $r \neq s$). Now, $\det_n (c, h)$ vanishes at an infinite number of points $(c_m, h_{r,s}^{(m)})$ of the irreducible curve $\varphi_{r,s}(c, h) = 0$ for $n \geqslant rs$. Hence $\det_n (c, h)$ vanishes at all points of $\varphi_{r,s}(c, h) = 0$ for $n \geqslant rs$ and therefore is divisible by $\varphi_{r,s}(c, h)$ when $rs \leqslant n$. The proof of Lemma 8.6, and hence of the Kac determinant formula, is complete.

Remark 12.5. There have been several published proofs of the Kac determinant formula: see Feigin-Fuchs [1982], Thorn [1984] etc. The proof given here is simpler, more elegant and works in the super case as well (see Kac-Wakimoto [1986]).

Remark 12.6. Let $h = h_{r,s}^{(m)}$ be one of the members of the list (12.1a, b) and suppose that there is no h' from this list (for the same m) with $h' - h$ a positive integer. Then in (12.16), (12.17) and (12.20) we have equality. This follows from the fact that the lowest eigenspace of d appearing in (12.13) has dimension 1 ($\Delta_{m,n;k}^j = 1$ for $j = k^2$) so that if $V(c, h')$ were contained in $U_{r,s}^{(m)}$, $h' - h$ must be a positive integer (see the remarks after (12.15)). In that case, by our hypothesis, (c_m, h') does not belong to the discrete series (12.1a, b) and hence cannot be unitary by the theorem of Friedan-Qiu-Shenker (FQS) [1984; 1986]; however, by the argument of Proposition 3.1, $U_{r,s}^{(m)}$ is a direct sum of unitary representations of *Vir*. We conclude that equality holds in (12.16), (12.17) and (12.20). The above condition holds in most, but not all, cases. In particular, it holds for $m = 1, 2$.

12.5. On non-unitarity in the region $0 \leqslant c < 1, h \geqslant 0$.

Recall that, according to the FQS theorem, all points in the *critical region* $0 \leqslant c < 1, h \geqslant 0$ not belonging to the discrete series (12.1) correspond to non-unitary representations of *Vir*. This together with Propositions 3.5 and 8.2(a) and Theorem 12.1(a), gives a complete classification of unitary highest weight representations of *Vir*: either $c \geqslant 1$ and $h \geqslant 0$ or $(c, h) \in$ discrete series (12.1).

In the remainder of this last lecture we shall discuss non-unitarity in the critical region. We shall obtain here only some partial results, which are, however, most important for applications (as in Proposition 3.8 or Remark 12.6, for example). The proof of the general result is more involved. See Friedan-Qiu-Shenker [1986] or Langlands [1986].

We call the n-th *ghost number*, and denote it by $g_n(c, h)$, the number of negative eigenvalues of the matrix of the Hermitian contravariant form $\langle \cdot \mid \cdot \rangle$ restricted to the n-th level of a highest weight representation with highest weight (c, h). Note that $g_n(c, h)$ depends only on n, c and h (in particular, it is the same for $V(c, h)$ and $M(c, h)$) and that $V(c, h)$ is unitary if $g_n(c, h) = 0$ for all $n \in \mathbb{Z}_+$.

Lemma 12.2. If $h \geqslant 0$ and $n \in \mathbb{Z}_+$, then

$$g_n(c, h) \leqslant g_{n+1}(c, h) \ .$$

Proof. Let $e = d_1, \alpha = -2d_0, f = -d_{-1}$. These elements form a standard basis of an $\mathfrak{sl}_2$ subalgebra of *Vir*, which we denote by $\mathfrak{a}$. Consider $V(c, h)$ as a representation of $\mathfrak{a}$. Given $v \in V(c, h)_{h+n}^e$ (i.e. $\alpha(v) = -2(h + n)v, e(v) = 0$), we have (see e.g. the proof of Lemma 8.1):

$$\langle f^k(v) \mid f^k(v) \rangle = k!(-\lambda)(-\lambda+1)\ldots(-\lambda+k-1) , \qquad (12.21)$$

$$ef^k(v) = -k(k-1 + 2(h+n)) f^{k-1}(v) . \qquad (12.22)$$

It follows that unless $v \in V(c,h)_h$ and $h = 0$ (then $\mathbb{C}v$ is a-invariant), the sub-space $T_v = \sum_{k \geqslant 0} \mathbb{C} f^k(v)$ is a space of an irreducible representation of a whose intersection with each $V(c,h)_{h+N}$ is 1-dimensional for $N \geqslant n$ (by (12.22)). Also, by (12.21), the Hermitian form $\langle \cdot \mid \cdot \rangle$ restricted to T_v is positive definite, negative definite or zero according as $\langle v \mid v \rangle > 0, < 0$ or $= 0$ respectively.

Using the Casimir operator of a, it is easy to see that $V(c,h) = \sum_v T_v$. This completes the proof of the lemma. $\quad\blacksquare$

Proposition 12.2. Let

$$\mathcal{D}_j = \{ (c,h) \mid 0 \leqslant c < 1, h \geqslant 0 \text{ and } \varphi_{j,1}(c,h) < 0 \},$$

where $\varphi_{r,s}$ are defined in (8.13). Then for every $(c,h) \in \bigcup_{j \geqslant 2} \mathcal{D}_j$, the representation $V(c,h)$ is not unitary.

Proof. We prove by induction on $n \geqslant 2$ that for (c,h) from the region $\mathcal{D}^{(n)} = \bigcup_{j=2}^{n} \mathcal{D}_j$, one has: $g_n(c,h) \geqslant 1$. The case $n = 2$ has been discussed in Lecture 8. Suppose that the statement is true for $n-1$. Then $g_n(c,h) \geqslant 1$ in $\mathcal{D}^{(n-1)}$ by Lemma 12.2. Let $(c,h) \in \mathcal{D}_n \backslash \mathcal{D}^{(n-1)}$; it is easy to see (using formulas (8.13) or (8.15)) that then $\varphi_{r,s}(c,h) > 0$ for all r, s such that $rs \leqslant n, (r,s) \neq (n,1)$. Hence, by the Kac determinant formula, $\det_n(c,h) < 0$, proving that $g_n(c,h) \geqslant 1$ in $\mathcal{D}^{(n)}$ except for the curve $\gamma = \mathcal{D}_n \cap \{ (c,h) \mid \varphi_{n-1,1}(c,h) = 0 \}$ (see Figure 12.1). The curve γ is tackled as follows. Recall that in the region $\mathcal{D}_n \cap \mathcal{D}_{n-1}$, $g_n(c,h) \geqslant 1$. But $\det_n(c,h) = \varphi_n(c,h)\varphi_{n-1}(c,h) \times$ (the rest) > 0 in this region since the first two factors are negative and the rest are easily seen to be positive. Hence we conclude that

$$g_n(c,h) \geqslant 2 \text{ for } (c,h) \in \mathcal{D}_n \cap \mathcal{D}_{n-1} . \qquad (12.23)$$

But the multiplicity of a zero of $\det_n(c,h)$ (as a polynomial in h with fixed c) is 1 for $(c,h) \in \gamma$. This shows that $g_n(c,h) \geqslant 1$ on γ. $\quad\blacksquare$

Corollary 12.1. If $V(\frac{1}{2}, h)$ is unitary, then $h = 0, 1/2,$ or $1/16$.

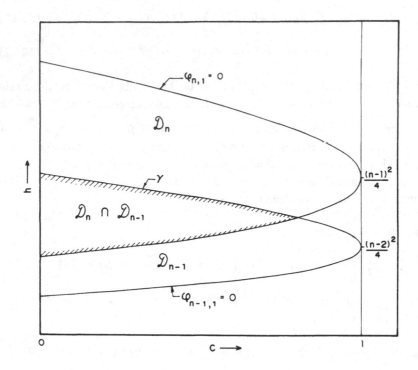

Figure 12.1

Proof. For $c = \frac{1}{2}$, Proposition 12.2 eliminates all points except for $h = \frac{1}{2}$ and $0 \leq h \leq 1/16$. But $\varphi_{2,2}(\frac{1}{2}, h) < 0$, and hence $\det_4(\frac{1}{2}, h) < 0$, for $0 < h < 1/16$. ∎

Remark 12.7. A similar argument shows that the unitarity of $V(c_m, h)$, where c_m is given by (12.1a) implies that $h = h_{r,s}^{(m)}$ given by (12.1b).

REFERENCES

Arbarello, E., De Concini, C., Kac, V.G., Procesi, C. [1987] "Moduli spaces of curves and representation theory", *Comm. Math. Phys.*, to appear.

Belavin, A.A., Polyakov, A.M., Zamolodchikov, A.B. [1984a] "Infinite conformal symmetry in two-dimensional quantum field theory", *Nuclear Phys.* **B241** (1984), 333–380.

Belavin, A.A., Polyakov, A.M., Zamolodchikov, A.B. [1984b] "Infinite conformal symmetry of critical fluctuations in two dimensions", *J. Statist. Phys.* **34** (1984), 763–774.

Benkart, G. [1986] A Kac-Moody bibliography and some related references, in: *Lie algebras and related topics*, CMS conference proceedings 5, 1986, 111–138.

Capelli, A., Itzykson, C., Zuber, J. B. [1987] "The A-D-E classification of minimal and $A_1^{(1)}$ conformal invariant theories", *Comm. Math. Phys.* **112** (1987).

Chari, V., Pressley, A. [1987] "Unitary representations of the Virasoro algebra and a conjecture of Kac", IAS preprint.

Chodos, A., Thorn, C.B. [1974] "Making the massless string massive", *Nuclear Phys.* **B72** (1974), 509–522.

Dashen, R., Frishman, Y. [1975] "Four-fermion interactions and scale invariance", *Phys. Rev.* **D11** (1975), 2781–2802.

Date, E., Jimbo, M., Kashiwara, M., Miwa, T. [1981] "Operator approach to the Kadomtsev-Petviashvili equation. Transformation groups for soliton equations III", *J. Phys. Soc. Japan* **50** (1981), 3806–3812.

Date, E., Jimbo, M., Kashiwara, M., Miwa, T. [1983] "Transformation groups for soliton equations", in: *Proceedings of RIMS Symposium*, M. Jimbo and T. Miwa, eds., World Scientific, 1983, 39—120.

Dirac, P.A.M. [1958] *The Principles of Quantum Mechanics*, 4th edition, Oxford University Press, 1958.

Feigin, B.L. [1984] "Semi-infinite homology of the Kac-Moody and Virasoro algebras", *Uspekhi Mat. Nauk* **39** (1984), 195—196. English translation: *Russian Math. Surveys* **39** (1984), No. 2, 155—156.

Feigin, B.L., Fuchs, D.B. [1982] "Invariant skew-symmetric differential operators on the line and Verma modules over the Virasoro algebra", *Funkt. Anal. i ego Prilozh.* **16** (1982), No. 2, 47—63, English translation: *Funct. Anal. Appl.* **16** (1982), 114—126.

Feigin, B.L., Fuchs, D.B. [1983a] "Verma modules over the Virasoro algebra", *Funkt. Anal. i ego Prilozh.* **17** (1983), No. 3, 91—92. English translation: *Funct. Anal. Appl.* **17** (1983), 241—242.

Feigin, B.L., Fuchs, D.B. [1983b] "Representations of the Virasoro algebra", Moscow University preprint.

Feigin, B.L., Fuchs, D.B. [1984] "Verma modules over the Virasoro algebra", in *Lecture Notes in Math.* **1060** (1984), 230—245.

Feingold, A. J., Lepowsky, J. [1978] "The Weyl-Kac character formula and power series identities", *Adv. in Math.* **29** (1978), 271—309.

Frenkel, I.B. [1981] "Two constructions of affine Lie algebra representations and boson-fermion correspondence in quantum field theory", *J. Funct. Anal.* **44** (1981), 259—327.

Frenkel, I.B., Garland, H., Zuckerman G. [1986], "Semi-infinite cohomology and string theory", *Proc. Nat. Acad. Sci. USA,* **83** (1986), 8442—8446.

Friedan, D., Qiu, Z., Shenker, S. [1985] "Conformal invariance, unitarity and two dimensional critical exponents", in *Publ. MSRI No.* **3** (1985), 419—449.

Friedan, D., Qiu Z., Shenker, S. [1986] "Details of the non-unitarity proof for highest weight representations of the Virasoro algebra", *Comm. Math. Phys.* **107** (1986), 535–542.

Fuchs, D.B. [1984] *Cohomology of infinite dimensional Lie algebras*, Nauka, Moscow, 1984; English Translation: Plenum Press, New York, 1986.

Garland, H. [1978] "The arithmetic theory of loop algebras", *J. Algebra* **53** (1978), 480–551.

Gelfand, I.M., Fuchs, D.B. [1968] "The cohomology of the Lie algebra of vector fields in a circle", *Funkt. Anal. i ego Prilozh.* **2** (1968), No. 4, 92–93 English translation: *Functional Anal. Appl.*, **2** (1968), 342–343.

Gepner, D., Witten, E. [1986] "String theory on group manifold", *Nucl. Phys.* **B278** (1986), 493–

Goddard, P., Kent, A., Olive, D. [1985] "Virasoro algebra and coset space models", *Phys. Lett.* **B152** (1985), 88–92.

Goddard, P., Kent. A., Olive, D. [1986] "Unitary representations of the Virasoro and super-Virasoro algebras", *Comm. Math. Physics* **103** (1986), 105–119.

Goddard, P., Olive, D. [1986] "Kac-Moody and Virasoro algebras in relation to quantum physics", *Internat. J. Mod. Phys.* **A1** (1986), 303–414.

Goodman, R., Wallach, N. [1985] "Projective unitary positive-energy representations of Diff(S^1)", *J. Funct. Anal.* **63** (1985), 299–321.

Jimbo, M., Miwa, T. [1983] "Solitons and infinite dimensional Lie algebras", *Publ. Res. Inst. Math. Sci.* **19** (1983), 943–1001.

Kac, V.G. [1974] "Infinite dimensional Lie algebras and Dedekind's η-function", *Funkt. Anal. i ego Prilozh.* **8** (1974), No. 1, 77–78. English translation: *Functional Anal. Appl.* **8** (1974), 68–70.

Kac, V.G. [1978] "Highest weight representations of infinite dimensional Lie algebras", in: *Proceedings of ICM*, 299–304, Helsinki, 1978.

Kac, V.G. [1979] "Contravariant form for the infinite-dimensional Lie algebras and superalgebras", in: *Lecture Notes in Physics* **94** (1979) 441–445.

Kac, V.G. [1982] "Some problems on infinite-dimensional Lie algebras and their representations", in: *Lecture Notes in Math.* **933** (1982), 117–126.

Kac, V.G. [1983] "Infinite dimensional Lie algebras", *Progress in Mathematics* **44**, Birkhäuser, Boston, 1983. Second edition, Cambridge University Press, 1985.

Kac, V.G., Kazhdan, D.A. [1979] "Structure of representations with highest weight of infinite-dimensional Lie algebras", *Adv. in Math.* **34** (1979), 97–108.

Kac, V.G., Kazhdan, D.A., Lepowsky J., Wilson, R.L. [1981] "Realization of the basic representations of the Euclidean Lie algebras", *Adv. in Math.* **42** (1981), 83–112.

Kac, V.G., Peterson, D.H. [1981] "Spin and wedge representations of infinite-dimensional Lie algebras and groups", *Proc. Nat. Acad. Sci. USA* **78** (1981), 3308–3312.

Kac, V.G., Peterson, D.H. [1984a] "Infinite dimensional Lie algebras, theta functions and modular forms", *Adv. in Math.* **53** (1984), 125–264

Kac, V.G., Peterson, D.H. [1984b] "Unitary structure in representations of infinite dimensional groups and a convexity theorem", *Invent. Math.* **76** (1984), 1–14

Kac, V.G., Peterson, D.H. [1986] "Lectures on the infinite wedge representation and the MKP hierarchy", in: *Séminaire de Math. Sup.* **102**, Montreal University, 1986, 141–184.

Kac, V.G., Wakimoto, M. [1986] "Unitarizable highest weight representations of the Virasoro, Neveu-Schwarz and Ramond algebras", in: *Lecture Notes in Physics,* **261** (1986), 345–371.

Kac, V.G., Wakimoto, M. [1987] "Modular and conformal invariance constraints in representation theory of affine algebras", *Adv. in Math.*, to appear.

Kaplansky, I., Santharoubane, L.J. [1985] "Harish-Chandra modules over the Virasoro algebra", in: *MSRI Publ.* **4** (1985), 217–231.

Kashiwara, M., Miwa, T. [1981] "The τ-function of the Kadomtsev-Petviashivili equation", *Proc. Japan Acad.* **57**A (1981), 342–347.

Langlands, R. [1986] "On unitary representations of the Virasoro algebra", IAS preprint.

Macdonald, I.G. [1979] *Symmetric functions and Hall polynomials*, Oxford University Press, 1979.

Sato, M. [1981] "Soliton equations as dynamical systems on infinite dimensional Grassmann manifolds", *RIMS Kokyoroku* **439** (1981), 30–46.

Segal, G. [1981] "Unitary representations of some infinite-dimensional groups", *Comm. Math. Phys.* **80** (1981), 301–342.

Skyrme, T.H.R. [1971] "Kinks and the Dirac equation," *J. Math. Phys.* **12** (1971), 1735–1743.

Sugawara, H. [1968] "A field theory of currents", *Phys. Rev.* **170** (1968), 1659–1662.

Thorn, C.B. [1984] "Computing the Kac determinant using dual model techniques and more about the no-ghost theorem", *Nuclear Phys.* **B248** (1984), 551–569.

Tsuchiya, A., Kanie, Y. [1986] "Unitary representations of the Virasoro algebra", *Duke Math. J.* **53** (1986), 1013–1046.

Virasoro, M.A. [1970] "Subsidiary conditions and ghosts in dual-resonance models", *Phys. Rev.* **D1** (1970), 2933–2936.

Wakimoto, M., Yamada, H. [1986] "The Fock representations of the Virasoro algebra and the Hirota equations of the modified KP hierarchy", *Hiroshima Math. J.* **16** (1986), 427–441.

For other references see Kac [1983], Goddard-Olive [1986], and Benkart [1986].